Energy modelling in architecture:
A practice guide

Energy modelling in architecture: A practice guide

Sonja Oliveira, Elena Marco and Bill Gething

RIBA Publishing

© RIBA Publishing, 2020

Published by RIBA Publishing, 66 Portland Place, London, W1B 1AD

ISBN 9781 85946 910 1

The rights of Sonja Dragojlovic-Oliveira, Bill Gething and Elena Marco to be identified as the Authors of this Work have been asserted in accordance with the Copyright, Designs and Patents Act 1988 sections 77 and 78.

All rights reserved. No part of this publication may be reproduced, stored in a retrieval system, or transmitted, in any form or by any means, electronic, mechanical, photocopying, recording or otherwise, without prior permission of the copyright owner.

British Library Cataloguing-in-Publication Data
A catalogue record for this book is available from the British Library.

Commissioning Editor: Ginny Mills
Assistant Editor: Clare Holloway
Production: Sarah-Louise Deazley
Designed by CHK Design
Typeset by Fakenham Prepress Solutions, Norfolk
Printed and bound by Short Run Press Limited, Exeter
Cover image: Tonkin Liu

While every effort has been made to check the accuracy and quality of the information given in this publication, neither the Author nor the Publisher accept any responsibility for the subsequent use of this information, for any errors or omissions that it may contain, or for any misunderstandings arising from it.

www.ribapublishing.com

Acknowledgements

First and foremost, we would like to express our gratitude to the many individuals across multiple architecture and design organizations for their inspiration, ideas and guidance that encouraged us to conceptualise this book and develop the topic. The general warm receptiveness and enthusiasm of all the contributors to the idea for this book was crucial in getting the chapters of this book going. We are also grateful for the support that we received throughout this project from our publishers as well as the anonymous reviewers for their encouraging and thoughtful suggestions.

    The lead editor Sonja would like to dedicate this book to her mum, Irene Dragojlovic, for whom she continues to have admiration, gratitude and love as she stoically perseveres with optimism and humour in self-isolation amid the Covid-19 crisis, as this book was going to press. This book would never have been accomplished without her support, patience, and encouragement.

	viii	About the editors
	ix	Case study contributors

An introduction to energy modelling
1

2 Introduction

6 **Chapter 1: Organisational responses to energy modelling in architecture**

Small firms
2

12 **Chapter 2: Prewett Bizley**
Testing and trialling PHPP

20 **Chapter 3: bere:architects**
Learning from the data and the detail of POE

26 **Chapter 4: Tonkin Liu**
Resource efficiencies and gains

Medium firms
3

36 **Chapter 5: KieranTimberlake**
Constructing a social energy modelling process

50 **Chapter 6: Henning Larsen**
Discovery and experimentation

64 **Chapter 7: Architype**
A question of culture

Large firms

4

82	**Chapter 8: AHMM** From programme to practice
96	**Chapter 9: Feilden Clegg Bradley Studios** The individual and the team
106	**Chapter 10: HOK** Achieving contextual design through a measured process

Conclusions

5

122	**Recommendations and conclusion**

124	References
126	Index
130	Image credits

About the editors

Sonja Oliveira is a sustainability and design innovation consultant and Associate Professor in Architecture at the University of the West of England, Bristol. Her practice is fostered through building strong industry, community, governmental and academic links across disciplines within the built environment as well as in computer science, environmental science and sociology. Sonja advises on innovative methods for delivery and dissemination of digital innovation and research strategies across leading architecture and engineering firms. She is a board member of the Serbian Green Building Council as well as Scientific Committee member of the Passive Low Energy Architecture (PLEA) network, HEA Fellow, institutional representative of the European Architecture Research Network (ARENA) and the Architectural Humanities Research Association (AHRA). Current ongoing projects include exploration of designers', educators' and students' imagined and lived experience of energy use in educational buildings, establishing a knowledge base to develop a residential building code for low energy housing across India and analysing residents' perceptions of energy efficiency measures in Bristol housing.

Bill Gething is a sustainability and architectural consultant, a Professor in Architecture at the University of the West of England, Bristol, a member of BRE Global's Governing Body and a CABE Built Environment Expert. Bill has broad design experience including master planning, bespoke office buildings, Higher Education buildings, housing and the design of caring communities built up over almost 30 years as a partner at Feilden Clegg Bradley Studios. His other consultancy work includes the development of the Green Overlay to the RIBA Plan of Work, work for the Energy Savings Trust on the thermal upgrading of hard to treat properties, leading the concept design and planning workgroup for the Zero Carbon Hub Design versus As Built Performance programme.

Elena Marco is an academic leader, educator, practitioner, and researcher, leading the third largest Department of Architecture and the Built Environment in the UK, at the University of the West of England (UWE) Bristol. She is a Principal Fellow of the Higher Education Academy (HEA) and has memberships of the RIBA New Courses Group, the RIBA Sustainable Futures Group, is part of the Council of the Standing Conference of Heads of School of Architecture (SCHOSA) and Member of the Council of Heads of Built Environment (CHOBE).

Case study contributors

bere:architects:
Justin Bere

Tonkin Liu:
Anna Liu

Prewett Bizley:
Robert Prewett

Architype:
Mark Lumley
Ann-Marie Fallon
Rebecca Robinson

Henning Larsen:
Jakob Strømann-Andersen (Henning Larsen)
Matthew Herman (BuroHappold)

KieranTimberlake:
Kit Elsworth
Billie Faircloth
Saeran Vasanthakumar
Ryan Welch

Feilden Clegg Bradley:
Joe Jack Williams

AHMM:
Craig Robertson

HOK:
Varun Kohli

An introduction to energy modelling

1

Introduction

Energy modelling in architecture has traditionally been the domain of the building services engineer, and typically has been used to verify the energy performance of a building design that is well advanced. At this stage, it is rarely possible to make any significant changes to the design without incurring major costs and delays, and opportunities to address fundamental flaws in the environmental aspects of the design will certainly have been missed. However, with recent advances in digital design technology, improved and more accessible simulation tools, and increasing emphasis on narrowing the energy performance gap, uptake of energy modelling tools by architects to inform early-stage design has been growing, both in the UK and internationally.

This book represents a compilation of early-adopter architecture firms' experiences, ideas and processes for integrating energy analysis and modelling across a range of scales of practice and project typologies. The book is not primarily about technology or the technical expertise needed to undertake energy analysis. Rather, it focuses on the social and organisational processes that might enable effective energy analysis and modelling integration in architecture practice.

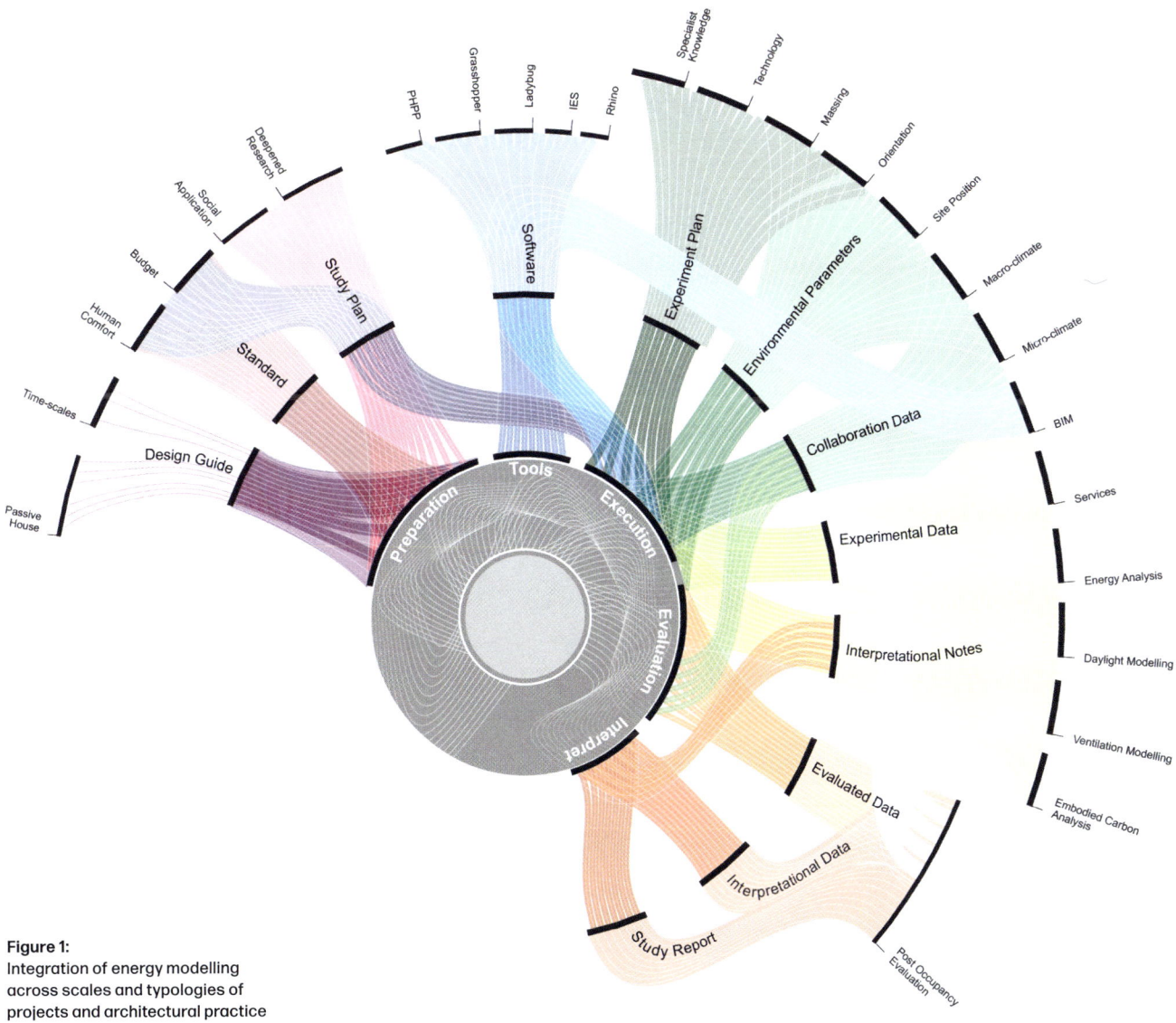

Figure 1:
Integration of energy modelling across scales and typologies of projects and architectural practice

Introduction 3

Embedding energy modelling into architecture practice is not straightforward, with early-adopter firms needing to develop new internal working arrangements and organisational structures, and new design methods. Many of the firms featured in this book found ways of integrating to vary depending on the type of project and also the scale of the practice (see Figure 1).

Chapter 1 introduces the findings from a research study into large UK architecture firms' approaches to integrating energy analysis and modelling across a range of projects. Parts 2, 3 and 4 of the book outline different-sized firms' approaches to integrating energy analysis and modelling, as viewed by the architects themselves. They recount different approaches to people, process and practice in the context of energy design in architecture.

Part 2 outlines the experiences of small firms, highlighting the importance of detail, personal stories and resource efficiency. For many of the small firms, there is an important evolution of skill, expertise and ability that is continuously documented, discussed and shared. Part 3 describes insights drawn from medium-sized firms, emphasising the importance of experimentation and relationship-building with design teams and other disciplines. Part 4 looks at the approaches of large firms, focused on developing and maintaining a consistent narrative across multiple office locations and teams. Here the importance of leadership across and between teams is highlighted.

Across all scales of firm and project, different organisational processes are highlighted and reflected on, including the importance of adjusting design workflows to include effective analysis, building and investing in appropriate resources within teams and firms, upskilling and learning how other design professionals including energy modellers might offer input, and experimenting with data and information across design cycles.

Finally, the conclusion explores potential future directions in both practice and higher education, highlighting the need for educational and practice approaches that build interdisciplinary social and organisational knowledge, skill and ability in addition to digital design expertise. The advancement of socio-organisational skill might allow for and enable not only analytical empathy for energy modelling, but other equally relevant and interconnected environmental, economic and social design needs.

We warmly thank the authors of the chapters for joining us in this undertaking. The combined contributions of the authors, editors, reviewers and others kept us continually aware of what a collaborative and collective social and organisational practice energy modelling in architecture is.

Chapter 1: Organisational responses to energy modelling in architecture

Energy analysis in architecture – whose role is it anyway?

Building services engineers have traditionally carried out energy analysis and modelling in building design. In their modelling work, emphasis is often placed on verifying established simulation models at late stages of design. However, calls for greater analytical input at the early stages of design mean that architects' involvement in energy modelling has become increasingly critical.

Recently, leading architectural firms in the UK and internationally have begun energy modelling processes at early stages of design within their practice, as evidenced in this book. Though there is widespread consensus that architects need to engage in these early energy modelling processes, the focus in practice, research and policy has been primarily on upskilling of technical expertise and knowledge.[1] Very often though, as illustrated in this book, the knowledge, ability and skill needs are not only technical, but social and organisational.

Chapter 2's reflections by Prewett Bizley suggest a need in small practice for social and organisational adaptation of the role of the architect to find ways of better understanding and analysing the 'energy' problems across projects. This better understanding involves development of bespoke tools to analyse specific 'energy' issues across projects. There was, as Prewett Bizley describe, 'no engineer to do the calculations'. Meanwhile Chapter 6's discussions by Henning Larsen suggest a need for organisational expansion of analytical skill drawn out of diverse disciplinary expertise, such as in-house research – a model that others in this book have adopted as a way to develop organisational analytical capacity. Their 'proactive search for outside knowledge' is described as a key 'defining element' of their approach to integration of energy modelling across projects. Chapter 9's description of an evolving approach by Feilden Clegg Bradley Studios similarly evokes a sense of trying to find what knowledge counts, and how to sustain and strengthen it in different teams dispersed across office locations.

In all the chapters in this book, questions of role pervade. The blurring of design need, skill and knowledge to enable analytical processes, often handed over to others, is continuously called on and reinforced. Broadening use of energy modelling to new user groups at early stages of design, and architects in particular, is viewed as a key way to enable improved analysis and prediction of building performance.

Though there are increasing calls for upskilling and retraining the UK construction workforce to meet increasingly stringent energy targets, how practice and higher education are responding remains largely undocumented. The following discussion begins to unpack some of the potential constraints to pathways being developed in large UK firms, drawn out of findings of a research study conducted by the editors.

Approaches to learning and sharing energy modelling outputs – findings from large UK firms

In recent studies conducted by the editors,[2] it was found that methods of learning, sharing and organising had diverse effects on whether, how and if energy modelling took place. In one of the studies, four large UK architecture firms participated in a series of focus group sessions and interviews, to understand how architects approached the implementation of energy modelling in their projects. The focus was to speak to a range of practice roles, including architectural technologists, project architects, and associate directors as well as directors in order to gather a range of experiences and approaches. In this instance, the energy modelling processes primarily referred to the implementation and potential use of Sefaira, cloud-based environmental building design and analysis software.

Interviews and focus group sessions were semi-structured and addressed the following themes: the role and background of a participant within the organisation, learning approaches to using the modelling tools, reasons for using the tools and methods for sharing 'modelling' knowledge with the client and building services engineer. Interviews and focus group sessions lasted between 45 and 60 minutes. A total of 52 participants took part across four large firms based in the UK.

When discussing learning approaches, most participants found the tool features easy to understand and learn (see Figure 1). For many, learning was organised through training sessions, but while these were found to be helpful, some participants conveyed their need to trial things outside of project fee-paying time. Learning overall was motivated by personal interest, rather than project or organisational need. 'Personal views are seen to largely support a moral, environmental and societal responsibility towards delivering considered design approaches.'[3] While learning was viewed as valuable, the application of acquired skill or knowledge was described as being difficult, dependent on others and often reliant on project needs. Client interests and needs as well as a firm's culture and approach to energy analysis were seen to be driving how and if energy modelling could be integrated.

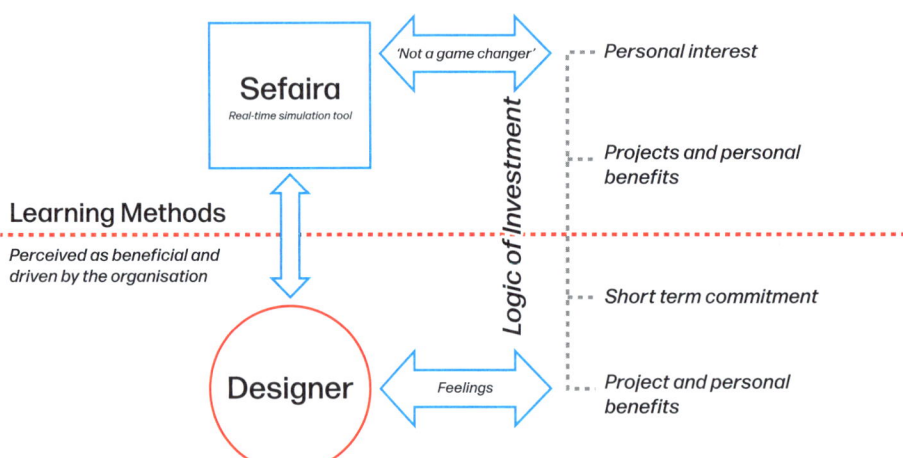

Figure 1: Learning approaches to integrating energy modelling in practice[4]

Where knowledge had been applied, the outputs of energy modelling and analysis were often discounted in favour of other design decisions such as aesthetics. Design responsibilities and sharing of outputs were seen as shrouded in potential liability risks. Participants, while recognising the need to engage and effectively apply energy analysis, would often point out that the responsibility for energy modelling outputs lay with building services engineers rather than architects. Current use of energy modelling as discussed within this study does suggest the development of a two-tier approach, whereby architects use the tools in-house to inform their design, while responsibility for the final modelling outputs lies with others in the design process.

Learning and application of knowledge, skill and ability can be bound by organisational and social issues. In the case of the firms that participated, these issues were mainly manifested through discussions of project and organisational interests and drivers, as well as notions of professionally bound responsibility, liabilities and risk – all of which constrained or enabled embedding of energy modelling in projects. In the case of this study, the findings reveal constraints. However, Parts 2 to 4 of this book discuss how the organisational and social context enabled an embedded approach, integrating energy modelling to effectively inform early design decisions.

The following section draws on experiences in higher education where an interdisciplinary curriculum design enables the blurring of the architect/engineer roles.

This suggests an approach by firms in this book that facilitates a potential pathway towards embedding integrated design in practice, and energy modelling in particular.

Experiences from higher education – interdisciplinary pathways

Most studies that have examined some aspects of energy efficiency teaching in higher education curricula have not focused on particular disciplinary approaches, nor have they included design professions. Rather, the emphasis has mostly been on measuring energy literacy and understanding students' energy-efficient behaviour in campus buildings.[5] Within built environment research there has been some work carried out to understand how sustainability-wide issues rather than specifically energy efficiency is taught, with discussions mostly focused on barriers to integration.[6]

Interdisciplinary learning is viewed by policy, practice and research as being a key way to enable integrated practice in the built environment. There is an established body of pedagogical built environment research that discusses potential pathways to inter- and multidisciplinary learning – through including sustainability content in a module or modules, or through enabling project-based learning or a more nuanced weaving of disciplinary learning across courses via events, lectures or projects. There are few examples that document or discuss integration of disciplines in an accredited course that includes both architecture and engineering.

In the UK, there are, to the authors' knowledge, three courses that offer integrated approaches to architecture and building services engineering. In UWE, The B(Eng) (Hons) Architecture and Environmental Engineering course is designed to enable graduates to gain an in-depth understanding, as well as the skill and ability to integrate analytical approaches drawn out of engineering and iterative reflective practice centred in architecture studio teaching.

In recent studies on the value of design studio as a mechanism of integrating disciplinary approaches,[7] it was found that the inherent fusing and weaving of architecture and engineering components (see Figure 2) in the course, as well as the social and organisational environment prevalent in studio, enabled learning, sharing and application of effective energy modelling in projects.

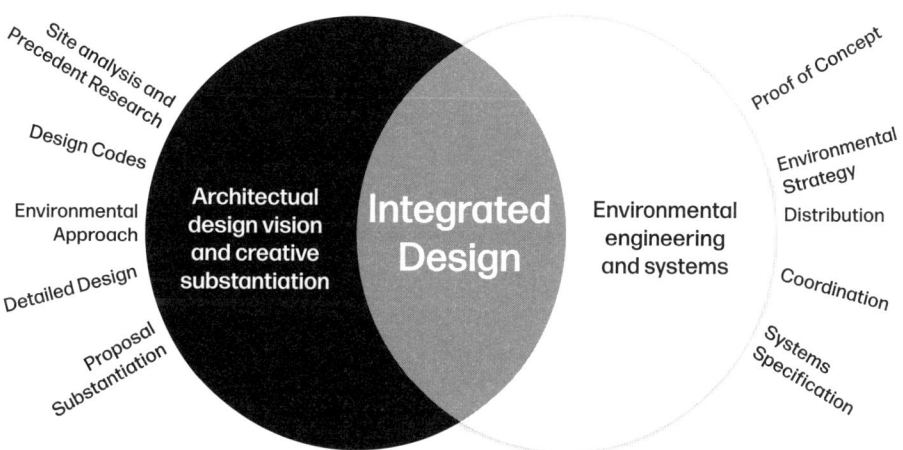

Figure 2: Integrating architecture and engineering in higher education

Intelligent design requires new multidisciplinary skills, and these need to be taught by professionals at the cutting edge of change. As noted by KieranTimberlake in Chapter 5, modelling practices are rituals and habits that might originate in one domain (engineering) and evolve over time to be the concern of another domain (architecture).

Small firms

2

Part 2 draws together case studies and reflections from three small firms: Prewett Bizley, bere:architects and Tonkin Liu. In all three, distinct approaches and experiences to integrating energy modelling are presented. **Prewett Bizley** reflect on the journey undertaken to enable an effective analytical approach using the Passive House Planning Package (PHPP), mostly in housing projects. They also draw attention to the importance of understanding detail and knowing how to interpret information to make intelligent design decisions, and make a case for informed analysis to drive good performance in housing. Meanwhile **bere:architects** discuss the importance of post-occupancy analysis and detailed understanding of submetered data as a way of distinguishing building performance from occupant usage. A detailed case study example of how monitoring of a housing project was achieved and how the data collected was interpreted is presented. Finally, **Tonkin Liu** reflect on a resource-efficiency driven approach, born out of a detailed and in-depth understanding of material properties and embodied energy.

Chapter 2: Prewett Bizley – testing and trialling PHPP

Thirty years ago, the construction industry was starting to wrestle with the idea of what sustainability and environmental design were. Energy modelling in the UK was still in its infancy, and the domain of very few rather technical people – and almost no architects. Architecture students were generally taught how to calculate U-values, but not given much information on what to do with them.

Without any access to modelling, the notion of what constituted low-energy design was extremely blurry, though it often seemed to involve sophisticated section diagrams with red and blue arrows indicating the flow paths of air currents. It was often unclear whether the colour-coded flow diagrams represented a clear understanding of how building physics really worked, or simply conveyed the desire of the designer. Irrespective of whether these vectors were correct, there seemed to be no way of quantifying magnitude.

Forced development of tools

In the early 1990s, Prewett Bizley emerged as a small architecture practice working mainly on one-off housing projects in the UK. It became increasingly apparent that existing housing in the UK accounted for almost 30% of carbon emissions, with heating contributing to over half of that. Housing projects could rarely afford an environmental engineer. It became clear to the practice that small architectural firms had the opportunity to a make significant impact in helping reduce carbon emissions, by encouraging deep low-energy retrofit approaches. Though the opportunity was there, the lack of early engineering input presented an obvious challenge.

The UK Standard Assesment Procedure (SAP) enabled architects to assess energy demand in the domestic sector. This was particularly empowering for architects, who were now able to observe where the energy in a household was going, and how the actions of the designer affected that. For the first time, designers were able to assess how alternative archtiectural approaches could lead to very different outcomes in terms of energy use. Architects were also, for the first time, being actively encouraged to use energy assessment as a way of informing design decisions.

This involved assessing:

- form, and its impact on heat loss
- window size, specification and fenestration pattern
- insulation type and thickness
- cold bridging, and how to avoid and mitigate it
- airtightness
- ventilation options.

Building on increasing use of SAP, the practice completed an extremely low-energy retrofit project in 2008. Monitoring of the project has shown it to perform almost exactly as predicted, and it remains one of the lowest-energy homes in the UK. This would not have been possible had the practice not been conversant with and in control of the energy modelling tools. As a practice, upskilling of designers meant design options could be tested quickly without having to prepare information, issue it to a third party and await a response. Having the agility and responsiveness to carry out in-house modelling was found to be very liberating.

Current practice

Twelve years after first trying SAP, the design process at Prewett Bizley has changed fundamentally. Now, for all projects, architects in the practice carry out energy assessments at the feasibility stage, using the Passive House Planning Package (PHPP). The practice tends to specialise in existing private homes, though they have also taken on social housing projects, and occasional newbuild housing projects too. The work on existing houses varies enormously, as each project comes with its own set of very particular constraints and opportunities. The variation occurs in the very construction system, and in the degree of planning restriction, and this often leads to unexpected outcomes in terms of energy use and architectural expression.

A case in point is the building at top left (see Figure 1). It is a Grade II listed townhouse from 1800. While the architects were faced with significant restrictions in how they could intervene with the original building fabric, they discovered that the form factor was so advantageous that by adding relatively modest amounts of insulation and upgrading windows with a special secondary system, they could transform the heating demand to near Passive House Standard.

Conversely, on another home (the 'semi' at bottom left in Figure 1), the form factor was much less favourable, and the architects needed to add much more insulation and yet still struggled to reduce the heating demand to Passive House Standard. However, in this case the planning system placed very little restriction on what energy efficiency measures the design team could employ.

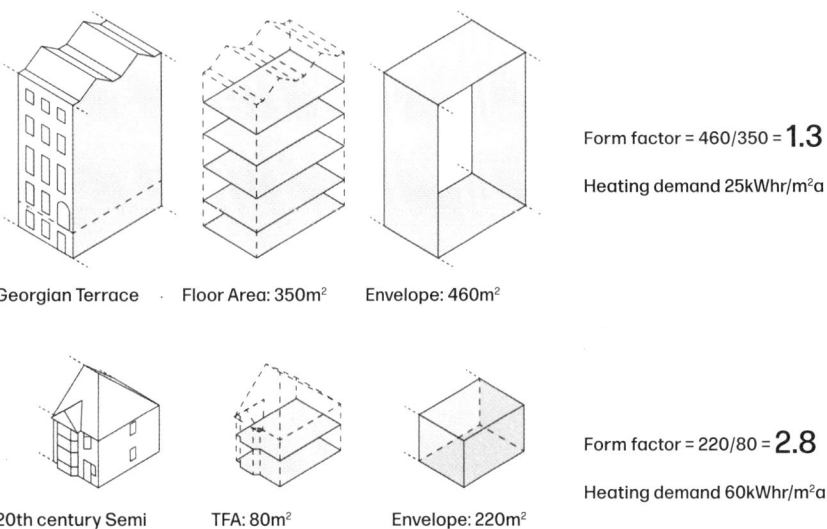

Figure 1:
Form factor comparision: buildings at the higher end of the form factor spectrum tend to be very hard to make energy efficient

Drawing on an evolving portfolio of work and a developing energy agenda in the practice, design processes can be characterised by two aspects:

- carrying out a careful energy study using PHPP software at feasibility study stage
- considering comfort at the same time as energy use.

Getting in early and doing some modelling at the beginning of a project is essential, as it means that designers can consider the implications of different design approaches at the start, which invariably is where the crucial design decisions are made. Leaving it later reduces the impact that modelling can have. Building energy models at the feasibility stage is critical, to determine what is possible, at what cost and within which constraints. The studies Prewett Bizley carry out include:

- energy consumption, focusing on specific space heat demand
- thermal bridging analysis.

Figure 2:
Annual heating demand for a house, modelled before and after retrofit

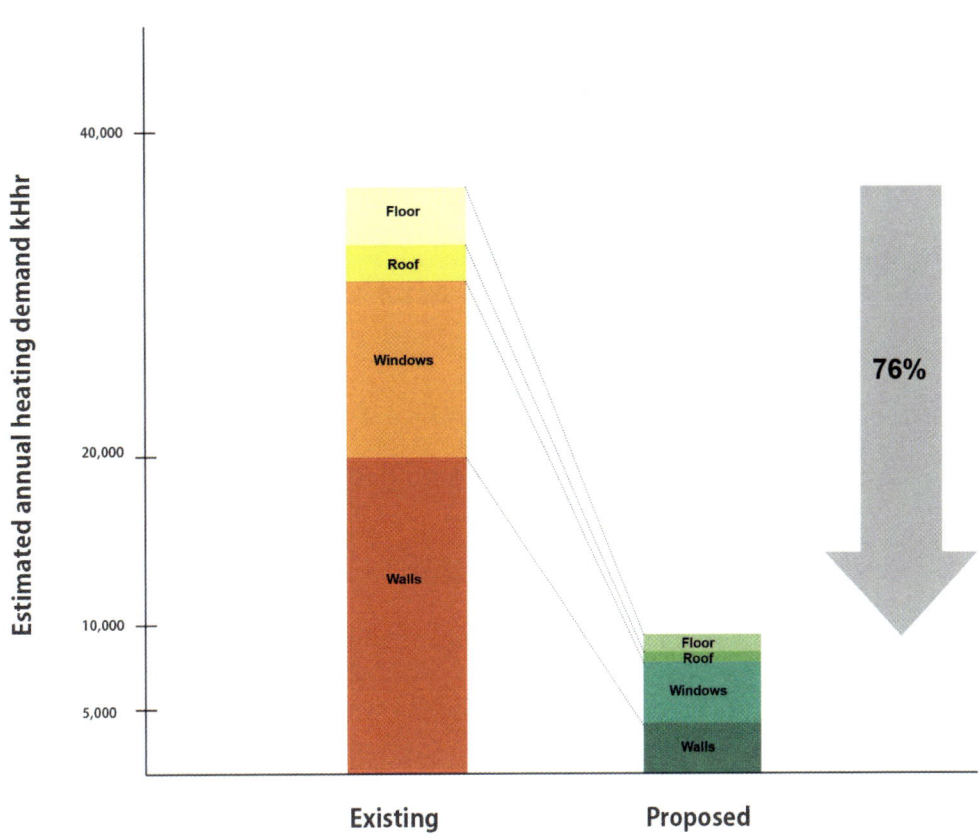

The diagram above is the sort of simple graphic Prewett Bizley might prepare during a feasiblity study to summarise the output from space heating comparisons between existing and proposed scenarios. As well as showing a headline heating reduction, the two bar charts indicate how effective fabric efficiency measures are expected to be for different building elements.

The notion of thermal bridging is familiar to most architects, but with increasing levels of insulation the impact of bridges is increasingly important, both in terms of heat loss and surface temperature. The complexities inherent in many thermal bridges makes calculating their effects hard to model using simple maths. Software such as THERM enables designers to visualise the impact of thermal bridges and determine their contribution to heat loss.

The two models in Figure 3 show analysis of thermal bridging at a corner junction within a 1960s house. The wall is a cavity construction, where the cavity is not continuous at the corners due to the need to carry beams in this area. The upper two diagrams illustrate the impact of the discontinuity of cavity insulation in two ways. On the left-hand side it can be seen that the surface temperature on the inside face of the wall may fall as low as 14.3°C when the external temperature is zero. The coloured diagram on the right-hand side shows the temperature in the construction very graphically, and it is clear that the temperature drop is quite localised. Local cool zones such as these are often associated with increased condensation risk and possible mould growth. The diagrams below show the impact of adding internal wall insulation (IWI), how this smoothes the isotherm lines, and how the surface temperature internally is greatly improved.

Option 1 – External wall at corner – Cavity fill

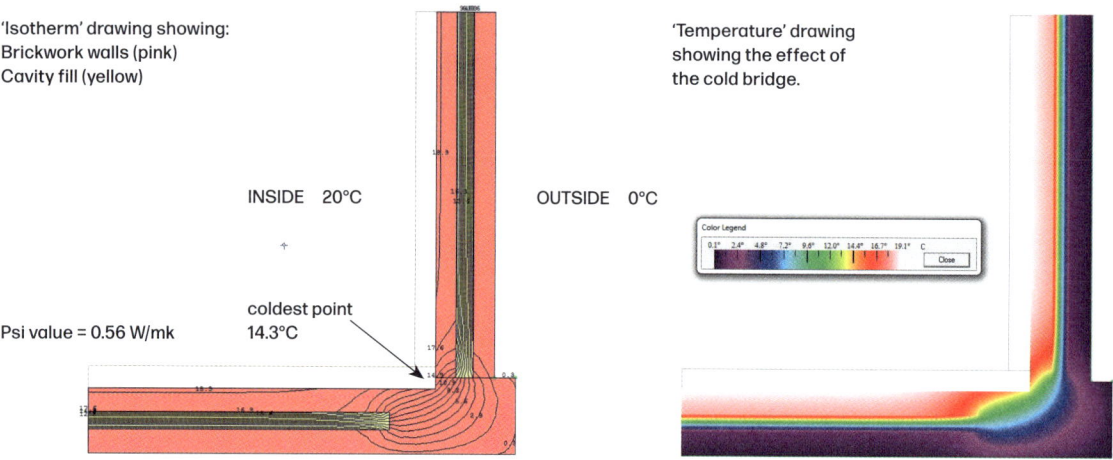

Option 2 – External wall at corner – Cavity fill plus IWI

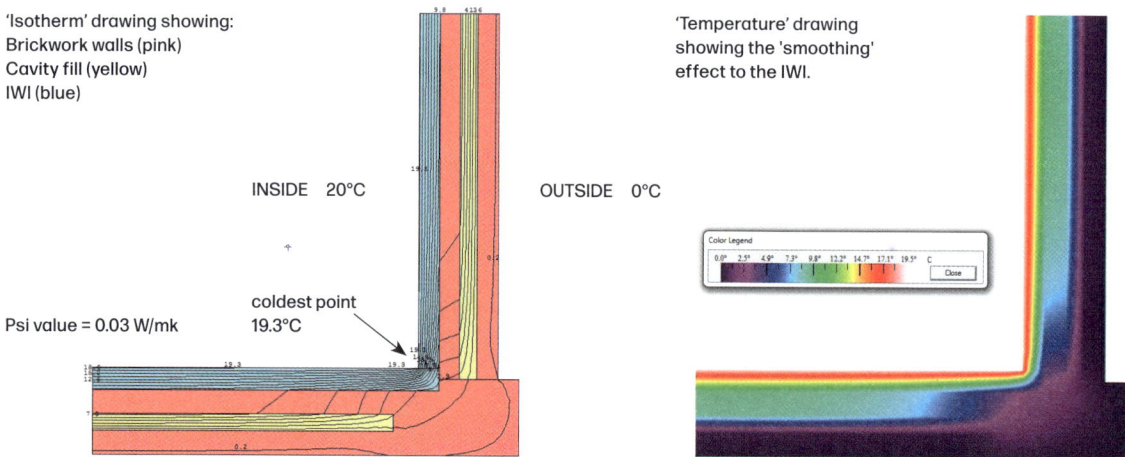

Figure 3: 'Isotherm' and 'temperature' drawings

As well as using modelling to control heat loss, Prewett Bizley increasingly test design work for overheating risks. PHPP predicts overheating risk by default, but the practice often tests a number of parameters in alternative models to see what the impacts are. The table in Figure 4 shows a number of options, with the percentage of time for which the space temperature exceeds a set threshold (such as 25, 26 or 27°C). Different specifications of glass and ventilation units are shown down the left-hand side, while across the top, different user behaviours have been modelled in terms of window opening.

Which software, and why?

Prewett Bizley have used PHPP for assessing heating and cooling demand, with a general aim of reducing the first and eliminating the need for the second through design. There are several benefits to PHPP, because it combines the following criteria:

Figure 4: Overheating scenario output table

		Scenario	Flat 1 (5 occupants) %	Flat 1 Hours	Flat 2 (5 occupants) %	Flat 2 Hours	Flat 3 (3 occupants) %	Flat 3 Hours	Flat 4 (2 occupants) %	Flat 4 Hours	Flat A (2 occupants) %	Flat A Hours	Flat B (3 occupants) %	Flat B Hours
Natural ventilation and MVHR	1	Natural Ventilation (Day&Night), Internal shutters, Low g-value glazing & MVHR with demand control	–	–	0.2	18	0.2	18	–	–	–	–	–	–
	2	Natural Ventilation (Day&Night)	0.1	9	2.3	201	2.6	228	0.1	9	0	0	0	0
	3	Daytime natural ventilation & Internal shutters	0.2	18	3.2	280	2	175	2.6	228	0	0	0	0
	4	Daytime natural ventilation & Low g-value glazing	1.1	96	3.8	333	3.1	272	3.3	289	0.2	18	–	–
	5	Daytime natural ventilation	1.5	131	5	438	4.1	359	5	438	0.8	70	0.1	9
MVHR only and windows not used	6	Windows closed, Low g-value glazing & MVHR with demand control	3.5	307	9.8	858	6.2	543	4	350	0.8	70	–	–
	7	Windows closed, Internal shutters, Low g-value glazing & MVHR	3.5	307	8.4	736	4.9	429	2.4	210	–	–	–	–
	8	Windows closed & MVHR with demand control	4.2	368	13.6	1191	8.4	736	6.8	596	1.8	158	0.1	–
	9	Windows closed, Low g-value glazing & MVHR	5.3	464	12.2	1069	7.9	692	4.4	385	0.9	79	0.1	9
	10	Windows closed & MVHR	6.1	534	16.5	1445	10.7	937	7.7	675	2	175	0.4	35

- it is relatively easy to use
- it is well understood
- it has plenty of design controls
- it makes testing quite easy
- it is affordable for small practices.

For thermal bridging Prewett Bizley usually assess specific junctions by eye, and for projects which are especially challenging or concerning they use THERM 5 computational modelling software. While the software is quite clunky, it is free (at the time of publication) and there are various guides produced by the Association for Environment Conscious Building (AECB) and others on how to use it successfully.

How do Prewett Bizley use the software?

There can be a fear among designers who use modelling software that either they will spend all their time modelling, or that the model will tell them how to design their project. In reality, neither is true, though planning in time for modelling and getting paid for it are obviously important.

Models do not design projects, though – designers do. What models do is provide information, such as how much energy and carbon a building is likely to consume, or how comfortable it is likely to be. Being aware of how design decisions affect these outcomes enables the designer to make better judgements. Sometimes the output of the model will run counter to another design parameter such planning constraints or architectural expression. Prewett Bizley have taken the view that they would rather know early and reach decisions with this knowledge, even if sometimes this is inconvenient. They don't feel that this is especially constraining or diminishes creativity; rather, it challenges designers to be more creative.

However, there may be occasions when a designer must weigh one criterion against another. Knowing the level of impact a choice may have on environmental criteria supports more nuanced decision-making, but it rarely leads to decision-making by default. Low energy and comfort should be considered essential prerequisites.

Does it work?

From Prewett Bizley's early attempts to build Passive House retrofit projects, they have found that the predictions of the software have been uncannily accurate at times. One such project was the practice's first deep retrofit project to a Victorian terrace in a conservation area (see Figure 5). It involved rebuilding parts of the house and insulating the entire building envelope with almost no visual impact from the outside.

Towards the end of the project, during the cold snap of 2010, the house was being decorated but the heating system was not yet commissioned. The architect arrived on site to find the painters in short sleeves while outside the snow had not thawed in a week. Their presence and that of a couple of halogen lights was enough to maintain around 20°C air temperature inside. While the calculations had predicted this, the architects

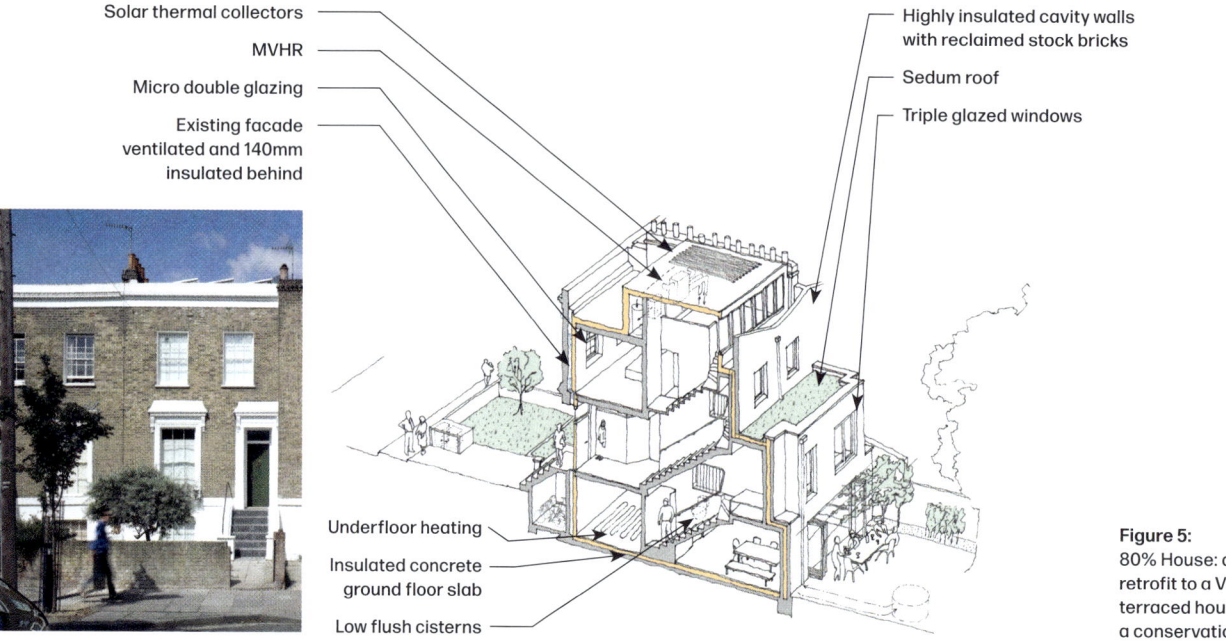

Figure 5:
80% House: a deep retrofit to a Victorian terraced house within a conservation area

were surprised by just how effective the works had been. That house has been monitored for eight years now, and the results are shown in Figure 6.

The graph in Figure 6 shows the gas usage by year since completion. There is very little variation year on year, and what variation there is appears to correlate with the severity of the winter. The red bar on the left shows the modelled gas consumption before the works, and the black dotted line shows the UK average. The actual gas use matches almost precisley predictions made by PHPP.

Prewett Bizley are collecting data on several other projects. The data collected reflects both energy use and levels of internal comfort such as temperature, relative humidity and carbon dioxide levels. There are three striking elements noted by the practice since they started adopting modelling to inform design:

- how great an effect it can have on outcome
- how easy it has been to adopt
- the comfort benefits.

There is a pervasive belief that making deep energy reductions in buildings is pointless, expensive or both. While it is difficult to suggest that low energy design is always

Figure 6:
Recorded gas consumption over eight years compared with estimated use before, and with the UK average

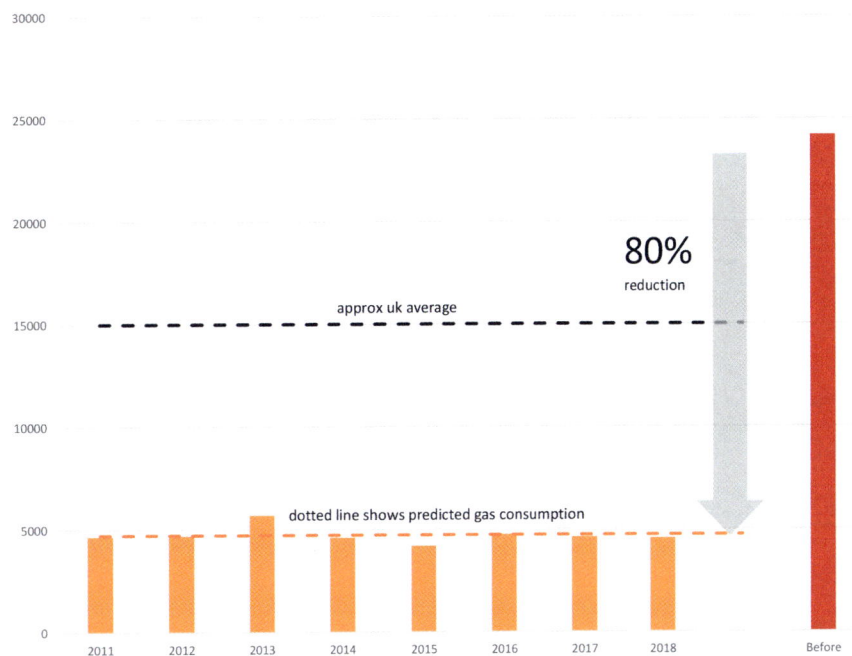

affordable, it is usually a relatively small part of the client's budget (say 10 to 20%, or to put it another way, the same as VAT). To reorganise one's house and employ an architect is by definition a costly enterprise. But for those who do it with at least a partial aim to become 'low energy', they can usually exceed the performance of new homes easily.

Using modelling has been revealing in terms of where heat energy goes. A big revelation is how important thermal bridging is when working with high levels of insulation. Common architectural features such as bay windows and dormers can be viewed in a rather different light. This is important in newbuild and retrofit design, as there is a tendency in architecture to overlook inconvenient truths. Having understood the heat loss/gain of a particular building, the designer can then control this in an informed, transparent way.

What next?

If the first stage in the evolution of Prewett Bizley has been adopting modelling, stage two has been and still is assessing the predictions by monitoring on site. So far heating demand and air quality seem to be showing good matches between prediction and monitoring.

Overheating is an increasing concern, in part because retrofit projects tend to have fixed window arrangements. Prewett Bizley are pushing back on this even in conservation areas and listed buildings, where they are using modelling and post-occupancy evaluation (POE) to justify ways of mitigating excessive solar gain. One recent success with this was gaining permission to fit external blinds within the dormer windows of a mansard of a Grade II listed building.

Prewett Bizley also use POE findings to have open and frank design conversations with clients, inviting them to consider carefully the consequences of fitting large windows without adequate shading. Managing overheating is also very much about developing an approach to better natural ventilation solutions, especially in urban areas. Opening windows simply is not possible for many people in the summer, as they are at work or asleep, and security demands trump the need to stay cool.

Chapter 3: bere:architects – learning from the data and the detail of POE

It is now relatively easy and inexpensive for architects to make meaningful energy performance comparisons against benchmarks, if they design their buildings in such a way that it is easy to gather sufficient data. This requires little more than appropriate electrical submetering, as well as the simple recording of metered data from any other fuel supplies. Combining harvested data with learning from post-occupancy evaluation (POE) interviews can help to achieve meaningful energy analysis that can be used to inform the early-stage modelling of subsequent projects.

Drawing on a number of case studies in this chapter, bere:architects offer a way through which POE can inform energy modelling early in design. In the mid-2000s, bere:architects collaborated with academic and industry partners to carry out in-depth research into the performance of some of the UK's first Passive House (PH) and near-PH domestic and non-domestic buildings. The research acted as a pathfinder programme, funded by the UK government's Technology Strategy Board (TSB) to find answers, with appropriate veracity, to questions of national importance. The TSB's construction research programme aimed to determine (1) where the biggest demand reduction opportunities lay, (2) whether this reduction could be achieved without affecting and perhaps even while improving the health and comfort of the occupants of buildings, and (3) the relative performance of different methods of design and construction in the UK climate. The programme first funded project research, then the TSB funded several meta-studies to compare project outcomes and answer some of the bigger questions mentioned above.

As a result of this research, in the UK we now have some useful benchmarks, enabling people to compare their own building's performance with other similar buildings, ranging from minimum-standard ones that meet the UK Building Regulations, to PH buildings that were found consistently to be top performers in terms of energy, health and comfort.

Learning from data

Using appropriate electrical submetering as well as records from metered data from any other fuel supplies, it is possible to assess and compare energy performance drawing on published benchmarks. Electrical submetering of circuits is a Building Regulations requirement for non-domestic buildings and is easily provided for within the fuse board. Wylex (a subsidiary of Siemens) and Hager are among those providing suitable electrical distribution boards containing submeters (see Figure 1).

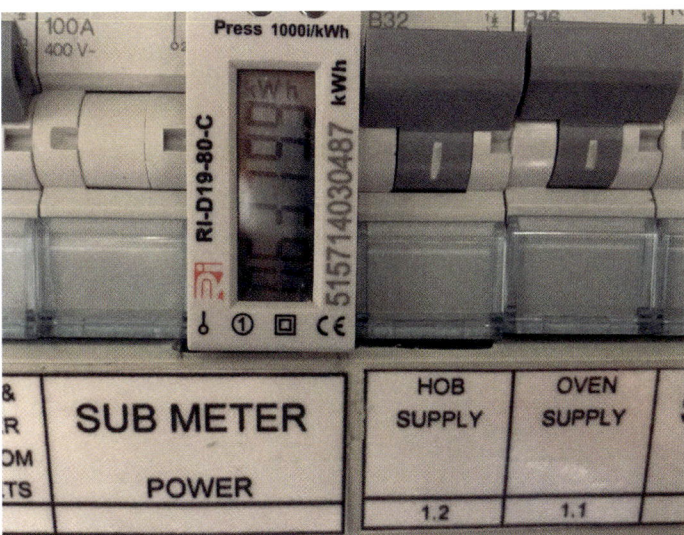

Figure 1:
Miscellaneous circuit from main distribution board at Lark Rise

If electrical submetering is provided for the following circuits, this will enable designers to separate 'building performance' from 'occupant usage', or what is often described as 'regulated' and 'unregulated' energy usage:

- heating (if electrically powered by a heat pump, for example)
- hot water (if electrically powered by a heat pump, for example)
- lighting
- cooking (if electrically powered, for example an induction hob)
- ventilation
- power sockets
- miscellaneous (pumps, etc.).

If periodic manual meter readings are taken, the data can be entered into an Excel spreadsheet, from which graphs are easily produced to make assessments:

- Energy consumption according to end-use category:
  - helping to understand the relative significance of contributory factors to overall building performance
  - helping to determine whether user habits are having the expected impact on performance.
- Energy use relative to design intent, national benchmarks and in-house projects:
  - overall energy use
  - submetered energy use.

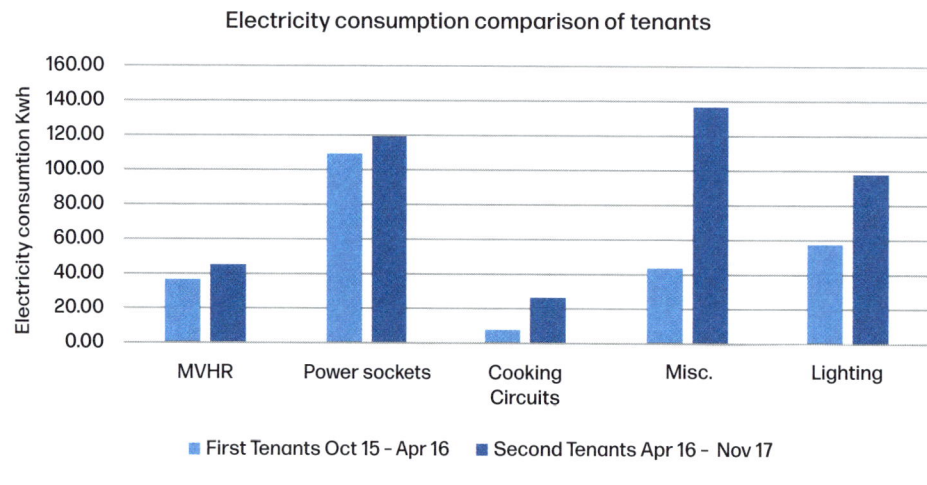

Figure 2: Graph showing energy use according to end-use category

Of course, all of the above end-use categories are affected by user behaviour. Occupant interviews can be highly rewarding in helping us to understand the reasons behind results, determining significant occupant factors:

- What is the usual set temperature for the heating, and is this often altered?
- How many baths or showers are taken each day, and what is the average duration of showers (domestic buildings)?
- What efforts are made to control lighting usage?

- What are the cooking habits (domestic buildings)?
- Unexpected factors - these can be significant. An outdoor hot tub and two bouncy castles were spotted on Google Earth at Lark Rise and occupant interviews confirmed these were left switched on for long periods in the summer. This explained the puzzle about the jump in miscellaneous energy use shown in Figure 2.
- Unregulated factors - what equipment is plugged into sockets, and how many appliances are left on standby?

Lark Rise

The value of understanding and interpreting submetering data is demonstrated by a POE carried out for a housing project, Lark Rise, in Buckinghamshire. Lark Rise is a detached, two-storey, two-bedroom dwelling located on a northwest-facing slope on the edge of the Chiltern Hills in Buckinghamshire.

Figure 3:
Lark Rise, north-west facing frontage

The energy consumption of power sockets, mechanical ventilation with heat recovery (MVHR), cooking, lighting and miscellaneous loads was monitored by submeters for an initial period of two years (October 2015 until October 2017). Lark Rise was designed to demonstrate how advanced-technology housing can provide heating and all other household needs including purification of its own waste products, while at the same time releasing existing grid electricity capacity and exporting excess energy to contribute to the growing demand for electricity for other uses. The purpose of this is to demonstrate the potential for a low-carbon transition plan that avoids demand shocks in the National Grid, releasing supply capacity for the electrification of cars and for the move towards heat pumps to decarbonise the heating of existing buildings.

The overall annual energy demand (kWh/m²/yr) of Lark Rise, analysed from data between October 2016 and September 2017, is lower than the PHPP design estimate, and more than eight times lower than a standard UK home of just 85m² (see Figure 4). Total consumption of heating and domestic hot water (DHW) was found to be below the PHPP design estimate, but electricity use for sockets, lighting, cooking and so on is higher than the PHPP design estimate.

Heating and DHW demands in the PHPP were calculated as 25% and 30% of the total demand of the house; however, they are in fact only 22% and 10% of the total demand of Lark Rise for the period between October 2016 and September 2017. This is a significantly smaller proportion of the total energy use than expected. Actual DHW consumption is particularly low, as this represents less than one-third of the PHPP design value.

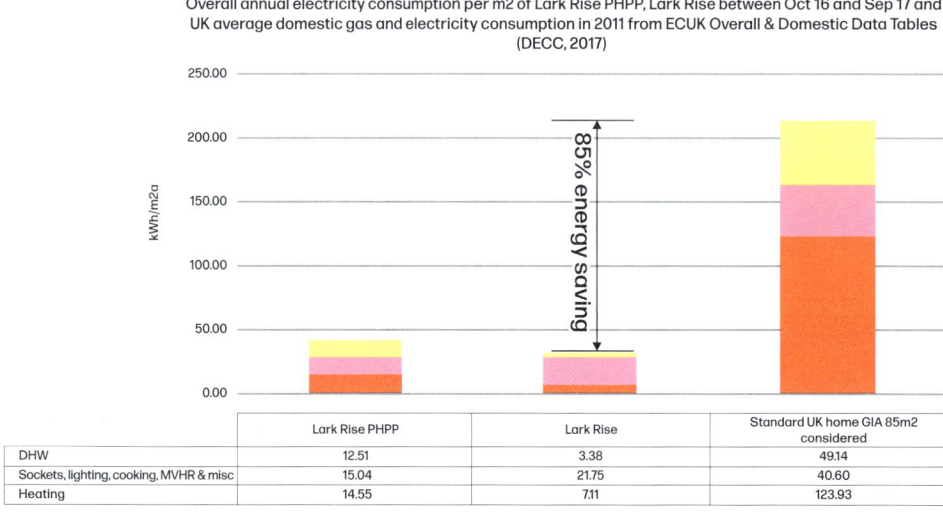

Figure 4: Comparison of overall energy use in Lark Rise against standard UK home energy use

The ventilation system's actual consumption is 9% of the total energy use at Lark Rise, which is twice as much as the 4% calculated in the PHPP. However, in October 2017 the authors of this report found that the intake and exhaust filters had not been changed since the unit was put into service two years earlier. The filters were clogged with black particulates in the intake filters (most likely from motor transport pollution) and white in the exhaust (dust from skin flakes, etc.). The poor state of the filters would certainly have created significant resistance to the flow of air, and this in turn would have significantly increased the fan power requirement. However, the actual consumption of the first tenancy period was also slightly higher (1.5 times) the PHPP design figure. It is thought that this could have been caused by blockage of the exhaust filter at an early stage, caused by dust in the air from decorating works. The authors of this report had specified that the ducts were to be stopped up until after completion; however, in practice this instruction was not properly complied with by the contractor.

Average consumption of power sockets and cooking (25% and 5%) represent a similar percentage of the total demand to the PHPP design figures (23% and 5%).

Miscellaneous demand (9%) is high compared to the PHPP design figure (2%). Google satellite pictures revealed the cause of this - an outdoor jacuzzi and two bouncy castles which one of the tenants admitted had been running 24 hours a day during the summer of 2017.

Lighting overall average demand, as shown in Figure 4, represents 19% of the total demand of the house; this is quite high compared to the 2% design estimate in the PHPP. Actual lighting design of the house (number of internal lights and quality) can affect this figure, but lighting consumption is also part of the floating demand of the house and therefore it can be affected by the lifestyle of the occupants.

Where next?

While Lark Rise shows how much useful data can be extracted by the simplest of means (five submeters) the recording and processing of the data is timeconsuming and hence can be costly in terms of labour. So this kind of monitoring will inevitably only be carried out by enthusiasts or researchers who have the opportunity to devote time to understanding the performance of an individual building.

However, smart controls are beginning to make it easier and more affordable to automate data collection and processing; thereby reducing labour costs and the risk of errors. In March 2020, bere:architects completed a privately-funded demonstration project; an optimally-orientated Passive House Plus home near Harpenden. This incredibly energy-efficient all-electric house is designed to be as non-polluting as possible in both construction and operation. Its structure is timber-framed, with recycled paper insulation, and it is equipped with rooftop power generation, battery storage, rainwater harvesting and sewage treatment. It has a bespoke Loxone control system which provides many user benefits and safety features, and also provides real-time monitoring, data storage and an onsite weather station.

The capital cost of such a controls and data monitoring system is high at present but could become relatively cheap and commonplace in a few years time. If this can be achieved, it will be possible to aggregate data from different buildings so that reliable summary statistics can be produced. Our aim with this project is to show that buildings can automatically and reliably generate and process performance data. Further, that the cost of such a system will become both affordable and essential if at scale it provides reliable benchmarking and the means to expose, regulate and control profligate energy consumption.

Chapter 4:
Tonkin Liu – resource efficiencies and gains

Tonkin Liu's energy modelling in architecture is grounded in a resource-efficiency first approach. This focuses attention on optimising structural design drawn out of material properties, as well as assessing the embodied energy in materials. Since their first lightweight prefabricated timber building, Camera and Jewel House, they have maintained a consistent approach of using a minimal amount of material to achieve the most. With the benefit of advanced digital tools, their design process that integrates space, structure and environment will continue to increase in precision, resulting in better integration and economy.

Shell lace structures

The practice's approach is manifested in a technique known as 'shell lace structure', pioneered by Tonkin Liu in 2009 with support from Arup engineers. This technique is characterised by a generative architectural approach that combines advanced digital tools with tailoring methods, informed by mollusc shells' formal structural principles. Tailored patterns of flat sheets are digitally generated and cut, then reassembled to form stiff, three-dimensional structures. From pavilions, to bridges, to towers, to large-span structures such as ferry terminals, all shell lace structure-informed designs have demonstrated astounding weight-to-span ratios.

Figure 1:
Mapping stress to porosity: Hull Solar Gate is 54% porous, spanning 10m height by 4m wide

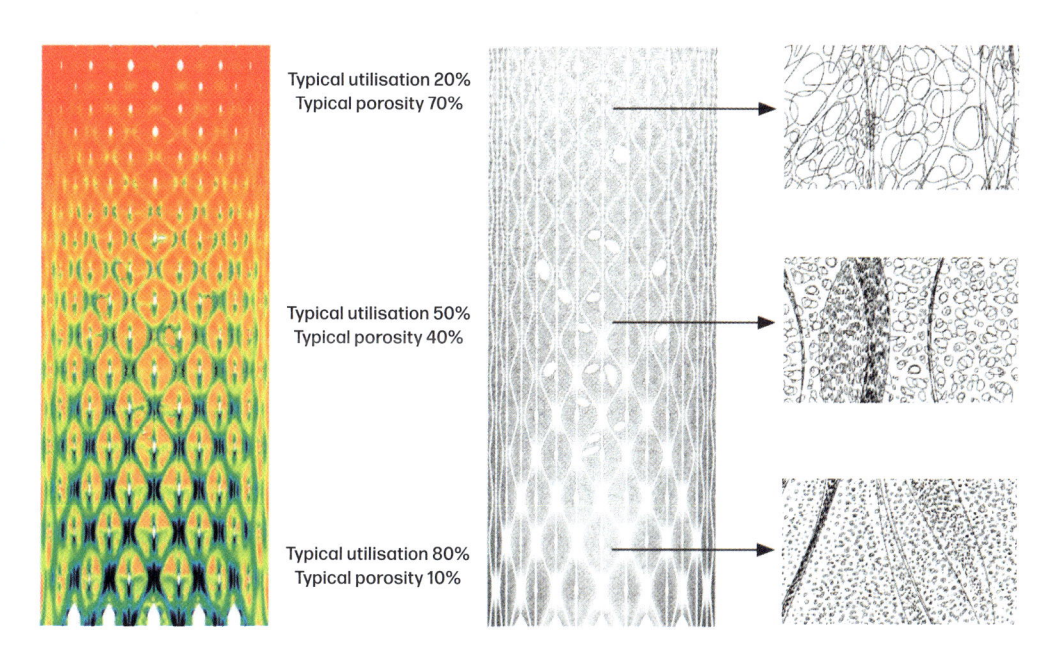

Typical utilisation 20%
Typical porosity 70%

Typical utilisation 50%
Typical porosity 40%

Typical utilisation 80%
Typical porosity 10%

The ten years of an evolving shell lace structure approach has produced for 18 ultra-lightweight engineered structures, six competition-winning designs, one RIBA research grant, one Innovate UK grant, one filed patent, three built projects and three projects currently in the pipeline. The resource-efficiency first approach came initially through necessity. For many architects in the UK, 2009 started a period of financial difficulty and an intense search for new work. Together with Ed Clark and Alex Reddihough at Arup, Tonkin Liu entered a flurry of design competitions. It was through the process of designing an entry for the RIBA competition for a seaside pavilion at Bexhill that the practice invented shell lace structure. Then, from 2009 onwards, they returned to it repeatedly, testing it with various scales, applications and details, and evolving what they felt to be a holistic architectural, structural, environmental and fabrication-driven approach to integrating energy analysis in design.

The context to this invention was two specific projects. In 2009, Tonkin Liu were delivering a large public-realm project in Dover alongside a penthouse in Marylebone, London. Using CNC-cut plywood moulds, the practice explored ways of casting the curvilinear precast concrete ramps, steps and walls for the Dover project. To create a set of environmental apertures for the penthouse, they learned about the increasingly affordable cost of perforating thin sheets of aluminium.

The design process for the Bexhilll pavilion began with a number of models, first in modelling clay and paper, then computer-generated models that were structurally analysed by engineers. The practice was endeavouring to find a way of cost-effectively making strong curved structures. Once the designers were happy with the macro and micro geometry, the computer model was flattened, laser-cut in paper and re-joined along the seams, like a tailored suit.

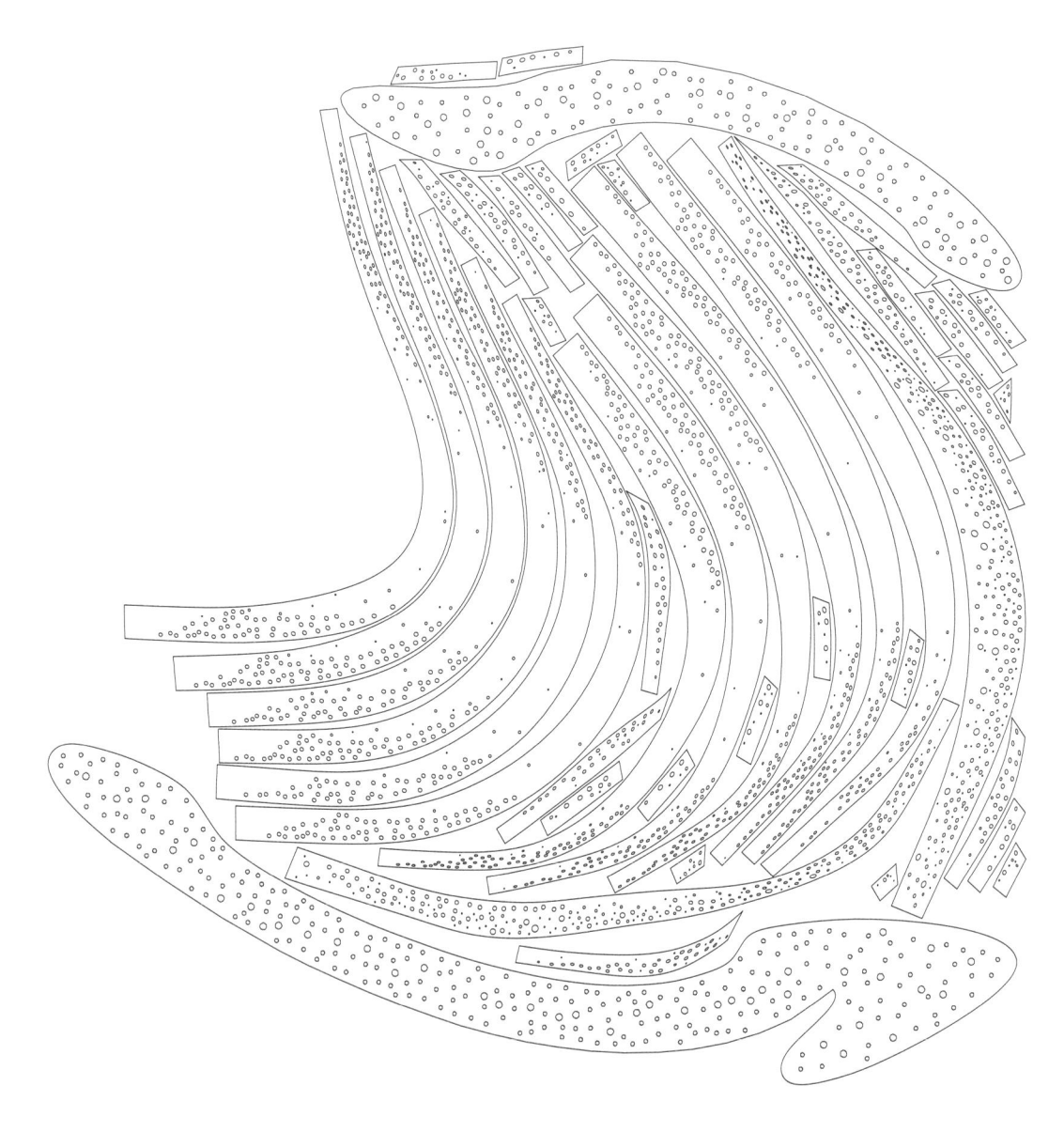

Figure 2:
Flattened computer model, to be sent for laser-cutting and re-joined like a tailored suit

Figure 3:
Tailored single-surface, lightweight and strong, energy-efficient, cost-effective structure – structural principles of Bexhill Pavilion by Tonkin Liu

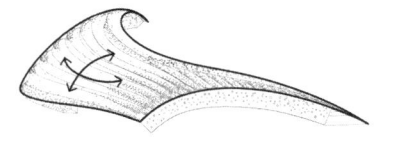

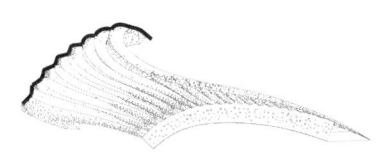

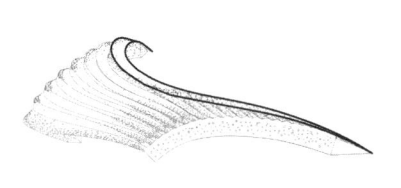

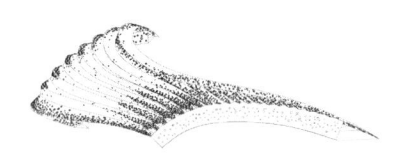

Curvature
The curvature of the shelter's structure gains strength through the resistance of its spherical and saddle shell forms and through the way the single surface structure spreads the loads in multiple directions.

Corrugation
Corrugation with an incline v-profile allows the surface to be constructed of flat sheets. The undulation of the corrugation transforms the surface into a series of hollow beams that vastly increase the spanning capabilities of the shelter's single surface structure.

Distortion
Distortion of the surface adds strength to both the shell form and the corrugation. The shell form is stiffened as it distorted between strong spherical and saddle forms. Distortion of the corrugation prevents the surfaces from concertinaing as they shift in orientation.

Stiffening
Torsion beams form benches inside and in front of the shelter that collect the ends of the tailored sheets. The closed triangularly form create a locked tailored ring beam that supports the entire shelter.

Perforation
Perforation lends transparency and lightness to the structure and reduces dead weight. The apertures display the thinness of the surface and by catching light, help the reading of the curving surfaces and tell a story of the flow of force.

Tonkin Liu had discovered a way of tailoring a single-surface, lightweight and strong, energy-efficient and cost-effective structure. This was the technique for which they would coin the term 'shell lace structure', which Ed Clark, a director at Arup, deemed 'a new breed of single-surface structure'. The economy in shell lace structure relies on all of the pieces being developable, meaning that they can be flattened. Therefore, only curves that are curving in one direction would work for shell lace structures – see Figure 3.

Nature's lesson on economy – form equates to strength
Nature-informed designs are not only formally sensuous, but also economical and energy-resource efficient. By necessity, the exuberant array of geometries found in biological forms employ a minimal amount of material to perform the maximum: lighter, bigger, longer, faster, stronger. Through observing the curvilinear geometry and details of mollusc shells and through an iterative process of 3D modelling and testing, Tonkin

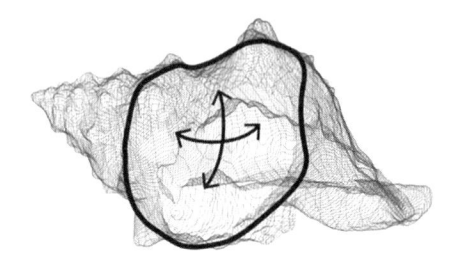

Curvature
Curvature lends strength to shell forms through the formal resistance of spherical and saddle forms and the ability of a single surface structures to deliver loads in multiple directions at the same time.

Corrugation
Corrugation of the surface transforms the plate structure into a series of hollow beams that vastly increase the spanning ability of the single surface and give direction to the applied forces.

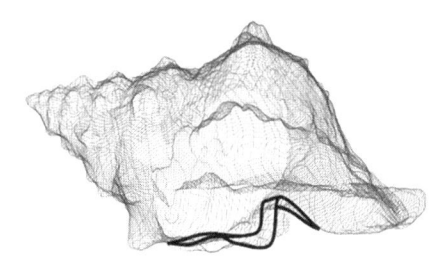

Distortion
Distortion of the surface adds strength to both the shell form and the corrugation by locking in stiffness and distributing the forces in different directions across the surface.

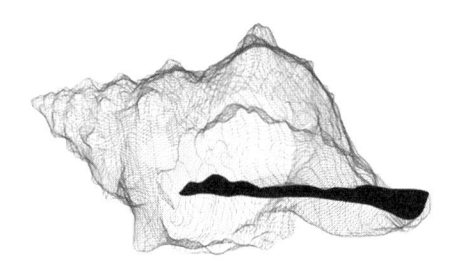

Stiffening
Torsion beams are created where the shells surface folds back on itself to make a hollow beam that often acts as a ring beam or core.

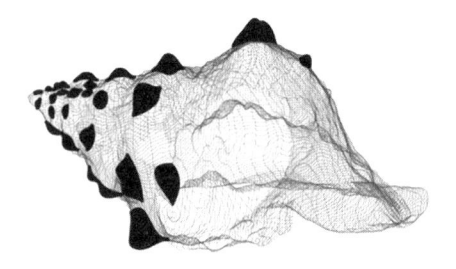

Nodules
Nodules rise up from the spherical form of the shell and the corrugation to add local stiffness that gives additional strength to the adjacent surface.

Figure 4:
Structural principles of mollusc shells by Tonkin Liu

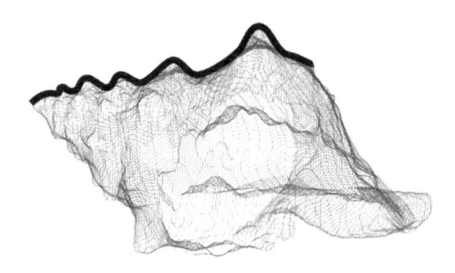

Liu discovered some principles in curved forms that contribute to strength. Curvature, corrugation, distortion, edge-stiffening, localised bending – all work together to achieve macro strength and local stiffness (see Figure 4). These can be considered as work-in-progress in a grand 500-million-year-long evolutionary design project.

Formal optimisation of shell lace structures

Shell lace structures would not have been possible or affordable ten years ago. Recent advancements in digital tools – modelling, analysis and fabrication – have enabled the practice to learn more from nature's constructional patterns and mimic its principles. This has been Tonkin Liu's interest since teaching at the AA School of Architecture from 2001 to 2005, and they have advanced deep into the territory of bio-mimicry. As a result, architects have an increasing understanding and control of the relationship between form and strength. Designers can now create and build structures of outstanding lightness and strength at relatively low cost. Macro and micro geometry directly translate into structural strength in ordinary sheet materials, enabling more to be built with less.

Optimising form equates to increased strength, which enables the shell lace structures to be constructed out of ultra-thin sheet materials. This cuts down on the consumption of material tonnage, and therefore the embodied energy. Large-scale prototypes have been fabricated for testing in workshop conditions. Two comparative analyses can be undertaken. Firstly, shell lace structures' comparative structural performance can be assessed, when each of the formal characteristics is taken away in turn: curvature, corrugation, twist. In doing so, designers can directly relate a formal characteristic to a particular structural characteristic. Secondly, designers can assess historic precedents of vaulted structures, for example by Dieste, Gaudí and Nervi, and quantify material tonnage to structural span ratio. This can be done by digitally creating shell lace structure versions of some of these precedents, and conducting structural analysis of these digital models. In doing so, it's possible to explore how shell lace structures, acting as tailored single-surface structures, can span volumes comparable to vaulted structures constructed in a very different way in the past, using a fraction of the materials. Digital tools are providing designers with the ability to manipulate a vast range of macro and micro geometries, perpetually responsive to the digital structural analysis. This results in economy both in the superstructure and the foundations, and highly tuned geometric structures that require a minimum of material to construct.

Shell lace structures' embodied energy

There are five main areas to consider when assessing shell lace structures' embodied energy: the material's sourcing, processing, fabrication and transportation to site, and energy for installation. First, the metal's sourcing – whether the material is being mined and rolled into sheets, or whether recycled steel sheets are being used. Second, for material processing, whether the designer is using unpainted or painted stainless steel, or mild steel that is then galvanised, depending on the environment of the site. Third, the shell lace structure's manufacturing should be considered, involving nesting the digital cutting pattern to minimise wastage, the laser-cutting, the positioning over digitally-cut formworks, then stitch-welding. Fourth, opportunities for nesting the pre-welded pieces to minimise size and cost of transportation vehicles, and the overall carbon cost of transportation and delivery should also be considered.

Finally, due to their inherent lightness, shell lace structures are supported on smaller foundations, and this involves lower energy and costs for lifting and installation, which should be taken into account when modelling embodied energy. Currently on site is the Manchester Tower of Light, a 40m tower built on the same principle, demonstrating ultra-lightness.

Figure 5:
Manchester Tower of Light stress plot by Arup

Why a resource-driven approach?

Tonkin Liu's shell lace structures journey embodies a dedication to a resource-efficiency first approach when considering energy modelling in design. It points to a greater consideration and better understanding of a material's structural, environmental and architectural properties as providing the initial 'energy DNA' of a design's possibilities. The discussion also highlights the important input that fabrication processes can have on testing the limits of an approach driven by efficiencies in resource extraction, manufacture, assembly and use.

32   Energy Modelling in Architecture

Figure 6:
Manchester Tower of Light by Tonkin Liu, due for completion in 2020; a highly tuned geometric structure requiring minimal material to construct

Medium firms

3

Part 3 collates insights and case studies from three medium-sized firms: KieranTimberlake, Henning Larsen and Architype. In all discussions, importance is placed on developing a mutually responsive engineering–architecture set of 'energy' relationships across projects, drawing on in-house research and careful documentation of what worked and what did not. **KieranTimberlake** discuss the importance of developing responsive energy workflows that provide effective feedback loops between energy analysis and design. **Henning Larsen** integrate BuroHappold's views and experiences into their discussion of developing a collaborative, multidisciplinary approach to energy modelling. **Architype's** section reflects on the practice's evolving knowledge base in the use of the Passive House Planning Package (PHPP) across non-domestic case studies.

All three firms highlight the need to understand integration of energy modelling as a socio-technical process in design, and not only a technically challenged one, calling on architects to engage in the definition of the energy model boundaries and analytical input.

Chapter 5: KieranTimberlake – constructing a social energy modelling process

Building energy modellers get into all sorts of trouble when others expect too much of their models. Building energy modelling is simply defined as the '*practice* of using computer-based simulation software to perform a detailed analysis of a building's energy use and energy-using systems'.[7] Based on this definition, building energy models should be the culmination of a simple series of input–output functions, a quasi-budgeting exercise, and therefore straightforward enough. However, the word *practice* belies the complexity.

Modelling *practices* are rituals and habits that might originate in one domain (engineering) and evolve over time to be the concern of another domain (architecture). Over the last ten years, energy modelling has become pervasive in the domain of performance-based architectural design, and there has been increased desire for, and scrutiny of, the use of models to test hypotheses about design features or to guide design to reduce building energy and improve comfort.[9] Energy modellers have concurrently matured their workflows but struggled to achieve a seamless and efficient feedback loop between analysis and design.

Evolution of energy model workflows at KieranTimberlake

Since 2010, KieranTimberlake have pursued an energy modelling workflow in tandem with their consulting engineers. This workflow is aimed at achieving a shared and co-authored energy model that allows for design iterations to keep up with the pace of design. The motivation for pursing a shared workflow has been shaped over time, and is informed by executing several architectural projects where the following observations have been made:

- Architects make better decisions when they conduct targeted studies using energy models during design.
- Architects do their best work when they engage and support energy modellers and their process.
- Model assumptions are important to architects and their clients, especially assumptions surrounding internal loads, system types and use schedules, because these inputs may significantly impact the relationship between architecture and energy performance.
- Architects can understand model assumptions, and can author and quality-assure the inputs related to their domain expertise.
- While models describe systems, modellers choose the system boundary.

Towards the beginning of this process, between 2010 and 2012, KieranTimberlake made several failed attempts to locate the origin of the issue by, for instance, attempting to execute model validation, verification and calibration routines, or by testing processes for design optioning with energy modelling software. While trying to locate the source of inefficiency and opportunity in the building energy model (BEM) workflow, the design team experienced the technical limitations of the software typically used by engineering consultants. Although the energy modelling process presented itself primarily as a technical challenge, through persistent collaboration, an understanding that the source of the problem is both technical and social evolved. The following sources of confusion and miscommunication often occurred:

- **Technical:** Energy modelling and architectural modelling require different levels of detail.
- **Technical:** Reconciling energy models and architectural models is possible but difficult.

- **Social:** The goal of and claims made about energy models vary, and are confusing.
- **Social:** Architects want to ask questions of energy models that they are simply not intended for.
- **Social:** Architects and energy modellers adhere to/do not question set system boundaries and default values.
- **Social:** Architects and clients may emphasise benchmarking instead of engaging studies that advance design.

The practice's position is now that a building energy model can be used as a design tool, but given the technical and social nature of the modelling process, it risks being marginalised as a method. Since 2012, KieranTimberlake have pursued a shared energy modelling workflow by working directly with consulting energy modellers. As early as the contract negotiation stage, the designers collectively define a modelling process that targets key parameters and prioritises sensitivity analysis, all within the context of the client's goals. Through building energy model kick-off sessions, there is a shared determination of modelling inputs, system boundaries and modelling timeline.

KieranTimberlake have also used their considerable in-house research resources to develop bespoke modelling workflows (see Figure 1) shaped through conversation with their consultants to address interoperability between architectural and energy models. Both architect and consulting engineer share responsibility for defining the model's system boundary, which has led to testing factors normally ignored in energy modelling practices.

The following section discusses the importance of establishing energy workflows when examining the role of landscape.

An expanded system boundary through vegetative modelling

KieranTimberlake's understanding of landscape's role in shaping microclimate and outdoor comfort is formed principally through first-hand experience: well-shaded green space offers welcoming respite from a hot summer sun. The same physical processes of absorption and evapotranspiration, which cool our bodies and temper the ambient environment, act on buildings as on people. Yet building energy modellers routinely – and often mandatorily – disregard these processes, and in doing so deprive architects of a key design tool for reducing energy use and improving thermal and visual comfort.

To examine how landscape and microclimate participate in the energy flows across a building envelope, the designer must address both the social and technical challenges presented by such ambitions. Performance-based paths for energy code compliance and certification schemes do not typically permit the inclusion of these features, though they do account for the shading context provided by permanent neighbouring structures. This professionally mandated reticence is understandable in light of energy modelling's dual purpose of system sizing and energy-use prediction: to design a cooling system which relies on trees that may lose their foliage, or die, is to run the risk of uncomfortable occupants. Yet in dismissing trees entirely from the heat-balance equation, one misses a valuable opportunity to incorporate a tuneable, seasonally selective shading device that is also a visual and ecological amenity. This prevents the design team from understanding the potential impact on HVAC, annual energy demand or costs. Pushing the industry to correct this omission will take significant time and advocacy. But in the short term, design teams should be encouraged to decouple considerations of compliance modelling requirements from studies and illustrate the value of performance-driven landscape design, developing standardised methods for dealing with the variability, uncertainty and geometric complexity of modelling tree shading.

The importance of considering vegetation as an energy model input became apparent to KieranTimberlake's team during an investigation of an existing masonry structure at the Philadelphia Navy Yard that would soon undergo a deep retrofit. While measuring the passive free-run temperature of the unoccupied building and its envelope, the design team observed a marked drop in the peak exterior wall surface temperatures between February and April, despite the shift to a warmer season. The source of this change became readily apparent when examining photographs taken just a few months apart, and it reveals what the designers knew intuitively: the dense foliage shades much of the masonry facade.

What the designers had not anticipated was the extent to which the elevated winter surface temperatures would produce a measurable trace through the entire wall assembly. An experiment in the subsequent winter involved heating the building to room temperature and allowing it to cool in order to understand the thermal diffusivity of the envelope. As the team tracked the temperature profile across the depth of the masonry wall over time, they discovered – much to their surprise – a temperature inversion, where the sunlight drove exterior surface temperatures above the inside temperature despite the cold February outside air. This effect would disappear in the presence of vegetation.

If the interaction of foliage and sunlight produced measurable effects on a masonry wall, then its effects on glazing ought to be far more significant. Hence, in subsequent projects KieranTimberlake have developed methods to estimate the cooling energy-saving potential of trees as a way of capturing their value to building projects. For example, for an unbuilt renovation-addition project at Tulane University, the team observed that several massive live oaks were well positioned to provide shading of the west facade.

Using techniques borrowed from forestry and ecology research, KieranTimberlake's designers estimated gap fractions and canopy volumes through on-site measurements that could easily be represented in 3D modelling software. They used these geometries and figures in solar studies to create hourly shading coefficients through simple insolation analysis for each portion of the building and, together with consultants at BuroHappold, translated their findings into hourly schedules that could be applied to each thermal zone and input directly into their energy model. These findings revealed a 19% average reduction in window solar gain, and as high as an 80% reduction in certain zones.

Figure 1:
Exterior surface temperature profile of a west-facing masonry wall in the Philadelphia Navy Yard

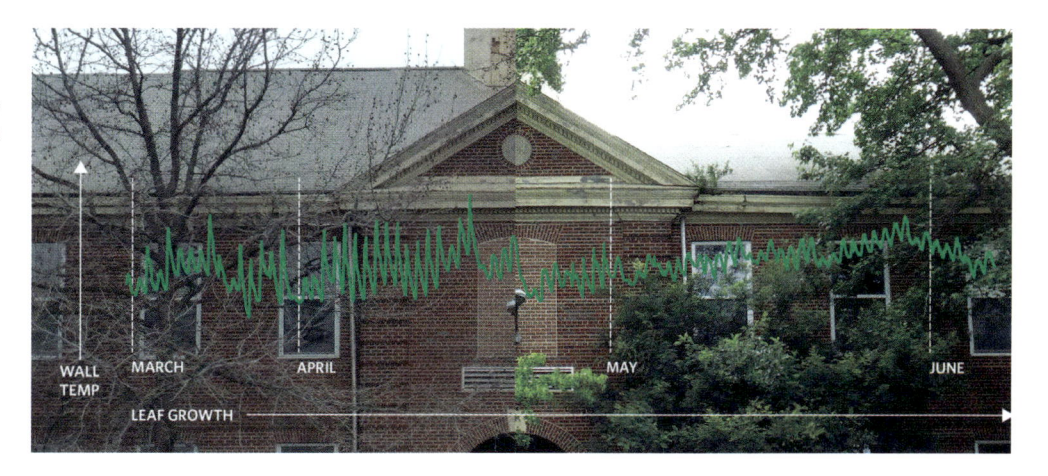

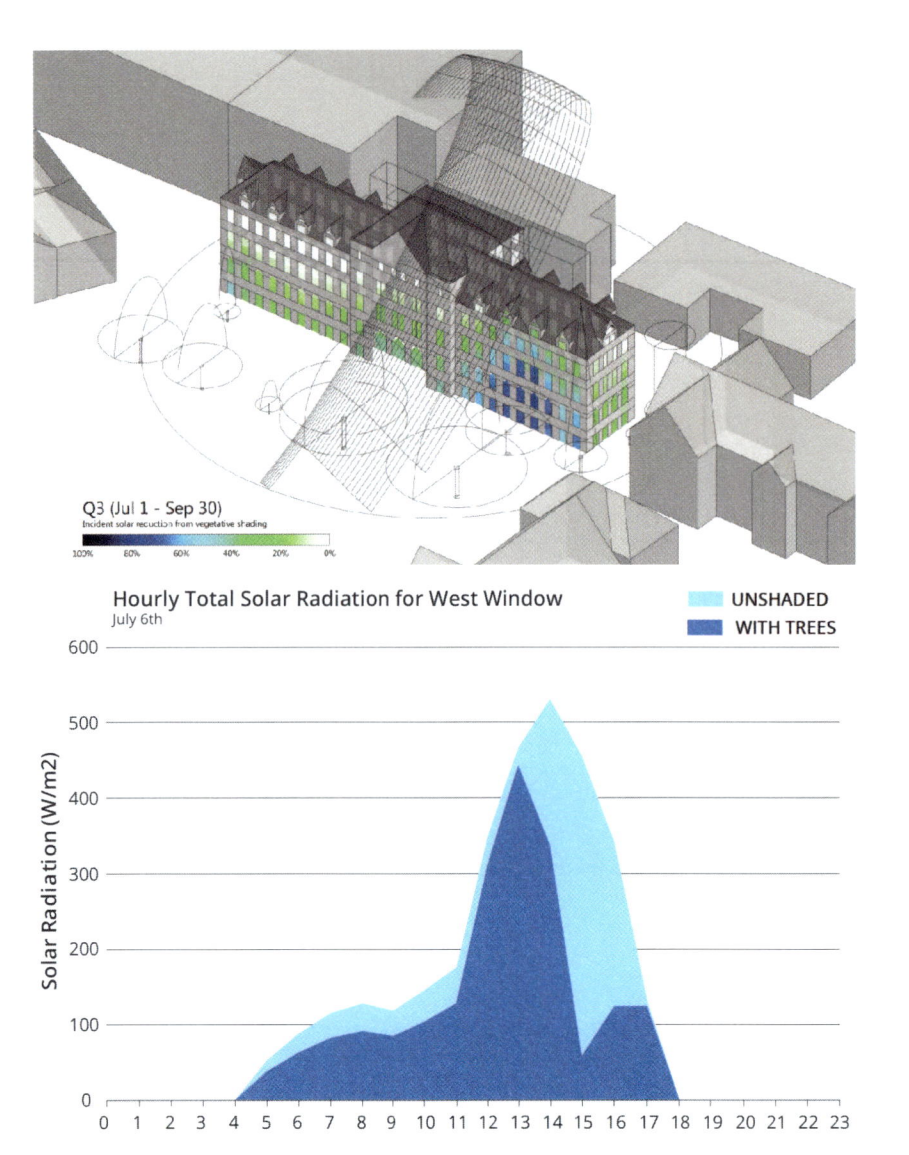

Figure 2:
For an unbuilt project at Tulane University, several massive live oaks were well positioned to provide shading of the west facade. KieranTimberlake developed models to estimate the cooling energy-saving potential of trees

By allowing the building energy model to account for the effects of site vegetation, KieranTimberlake effectively participated in defining the system boundary of the energy model. Although they were fortunate to possess the in-house horticultural expertise to measure the trees on site and translate those into digital geometries, this is not a discipline or skill exclusive to KieranTimberlake. This technical challenge can be replicated, through collaboration with students, researchers or consultants, to accurately account for vegetative shading on a project site. This example not only highlights the opportunity for architects to engage in the definition of the energy model, but it also allows the design team to include additional context in the analysis that will inform design decisions that may reduce energy and maintain comfort.

Architectural facade details – translating architectural shading into energy model inputs

These landscape performance studies have led designers at KieranTimberlake to consider how far they can extend the system boundary of the building energy models to consider context outside the scope of standard modelling practice. Even within the purview of standard practice there are meaningful departures between energy models and architectural models that can make it difficult for architects to fully engage energy modelling to guide their design work. Differences in model resolution between architectural details and coarser whole-building energy model representations frequently make it difficult to evaluate how finer architectural details, such as shading elements, facade depth and texturing, and interfaces between architectural assemblies, influence heating and cooling loads. Developing methods to bridge these divides can be essential for proving out hypotheses and capturing the greater value of these details in terms of building performance.

In a series of studies for a high-rise New York University project, KieranTimberlake explored ways to respond to solar exposure with high-performance glazing and self-shading features in a dense urban setting. Incident solar radiation studies of the lower six levels revealed highly varied exposure, due to the urban context and the rotation of the Manhattan street grid relative to true north. By pleating the unitised curtain wall in areas of highest exposure they were able to direct vision glazing away from the most intense sunlight and provide shading to adjacent units. These areas could also be further adapted with higher-performing low-emissivity coatings while leaving better-shaded areas available for more transparent glazing. While these studies were straightforward to execute using architectural solar analysis tools, evaluating their effect on cooling energy and peak cooling load proved far more difficult. KieranTimberlake found that the level of detail was far too high to use in a whole-building energy model, whose scale and complexity necessitated that facade geometry be represented through simple, flat surfaces.

Working closely with KieranTimberlake's environmental consultants and energy modellers at Atelier Ten, the team developed a workflow to capture the effective hourly reductions (or increases) in incident radiation due to the detailed geometry and condense this information in a series of monthly schedules for each thermal zone. This process involved comparing hourly incident radiation data for the actual pleated geometry against an equivalent set of studies for the simplified flattened geometry and captured the effects of detailed shading geometry in a numerical form that was easily adopted in the energy model. An unexpected benefit of this workflow was that with hourly incident radiation schedules for each region of the building envelope, the team could develop real-time estimates of window solar gains in response to localised changes in solar heat gain coefficient (SHGC). Using a spreadsheet representation of the building's unrolled elevations, the team could select from a range of glazing types for each curtain-wall unit and automatically compute the SHGC and consequent solar gains for the associated thermal zone as the product of SHGC and incident radiation. This allowed them to rapidly assess where improving glazing performance afforded the greatest potential cooling energy reductions, and ensured that the composition settled on before handing off shading schedules to Atelier Ten was already well informed. This process also allowed for rapid computation of effective SHGC and U-values for each thermal zone based on the distribution of selected glazing and unit types, and these figures were also used as inputs into the energy model.

This process was made possible through close collaboration with the energy modeller, who provided direction in determining how a shading schedule could be applied to facade elements in lieu of actually designing the pleated facade in the energy modelling software. KieranTimberlake's in-house efforts to unpack the energy model and spatially visualise the solar radiation in elevation helped the design team gain intuition and guided their glazing specification. Together, this process enabled the design team to shape the energy model in non-traditional ways, capturing shading of unique architectural details and authoring inputs into the energy model.

Figure 3:
Incident solar radiation studies of the lower six levels of a high-rise New York University project led KieranTimberlake to develop high-performance glazing and selfshading features.

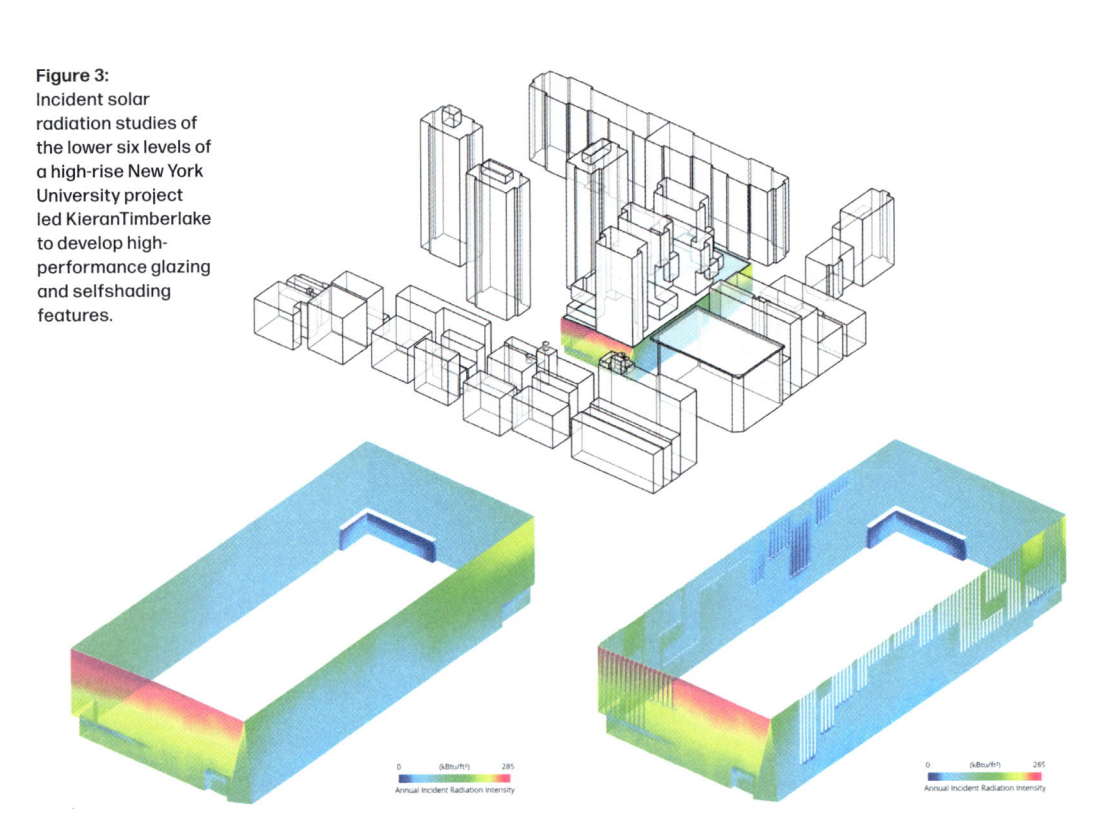

		Hourly 'projected' incident radiation intensity per façade (Btu/ft²hr) This value is determined by calculating the cumulative incident radiation on each façade at each our of the year and dividing the result by the area of an equivalent flat or 'projected' façade							Hourly average shading coefficient due exclusively to wedge panels This is found as the ratio of incident solar radiation intensity on a wedge panel façade as a fraction of equivalent incident solar radiation intensity on a flat façade		
		1.00	0.84	1.00	0.84	1.00	0.96	1.00			
Date/Time	Day Of Year	East Podium Façade - Flat	East Podium Façade - Wedge	West Podium Façade - Flat	West Podium Façade - Wedge	South Podium Façade - Flat	South Podium Façade - Wedge	North Podium Façade - Flat	East Podium Façade	West Podium Façade	South Podium Façade
Aug 01 01:00	Aug 01	0.00	0.00	0.00	0.00	0.00	0.00	0.00	1.0000	1.0000	1.0000
Aug 01 02:00	Aug 01	0.00	0.00	0.00	0.00	0.00	0.00	0.00	1.0000	1.0000	1.0000
Aug 01 03:00	Aug 01	0.00	0.00	0.00	0.00	0.00	0.00	0.00	1.0000	1.0000	1.0000
Aug 01 04:00	Aug 01	0.00	0.00	0.00	0.00	0.00	0.00	0.00	1.0000	1.0000	1.0000
Aug 01 05:00	Aug 01	7.73	6.69	4.89	4.18	6.21	5.91	6.12	0.8648	0.8538	0.9516
Aug 01 06:00	Aug 01	34.59	30.91	10.06	8.59	12.76	12.14	52.44	0.8937	0.8540	0.9516
Aug 01 07:00	Aug 01	53.45	47.18	13.98	11.94	17.73	16.87	70.33	0.8827	0.8540	0.9516
Aug 01 08:00	Aug 01	74.05	72.37	15.48	13.22	19.65	21.17	50.48	0.9774	0.8539	1.0772
Aug 01 09:00	Aug 01	89.87	85.15	17.43	14.88	25.50	26.09	19.92	0.9475	0.8538	1.0233
Aug 01 10:00	Aug 01	101.49	92.46	16.58	14.16	61.69	62.70	18.90	0.9110	0.8542	1.0164
Aug 01 11:00	Aug 01	66.89	53.63	15.86	13.56	96.43	92.66	17.99	0.8018	0.8549	0.9609
Aug 01 12:00	Aug 01	15.10	11.99	17.42	13.63	122.37	120.30	17.38	0.7945	0.7827	0.9831
Aug 01 13:00	Aug 01	14.57	11.57	69.24	53.97	134.98	128.92	16.83	0.7945	0.7795	0.9551
Aug 01 14:00	Aug 01	15.98	12.70	80.65	67.36	129.17	119.93	18.39	0.7945	0.8352	0.9285
Aug 01 15:00	Aug 01	15.10	11.99	65.99	57.91	108.22	97.21	17.38	0.7945	0.8776	0.8983
Aug 01 16:00	Aug 01	11.74	9.33	53.01	47.27	79.17	67.82	13.52	0.7945	0.8917	0.8567
Aug 01 17:00	Aug 01	8.56	6.80	49.74	46.68	42.38	33.97	9.90	0.7945	0.9384	0.8016
Aug 01 18:00	Aug 01	3.09	2.46	21.67	20.83	10.04	8.43	3.56	0.7945	0.9612	0.8396
Aug 01 19:00	Aug 01	0.00	0.00	0.00	0.00	0.00	0.00	0.00	1.0000	1.0000	1.0000
Aug 01 20:00	Aug 01	0.00	0.00	0.00	0.00	0.00	0.00	0.00	1.0000	1.0000	1.0000
Aug 01 21:00	Aug 01	0.00	0.00	0.00	0.00	0.00	0.00	0.00	1.0000	1.0000	1.0000
Aug 01 22:00	Aug 01	0.00	0.00	0.00	0.00	0.00	0.00	0.00	1.0000	1.0000	1.0000
Aug 01 23:00	Aug 01	0.00	0.00	0.00	0.00	0.00	0.00	0.00	1.0000	1.0000	1.0000

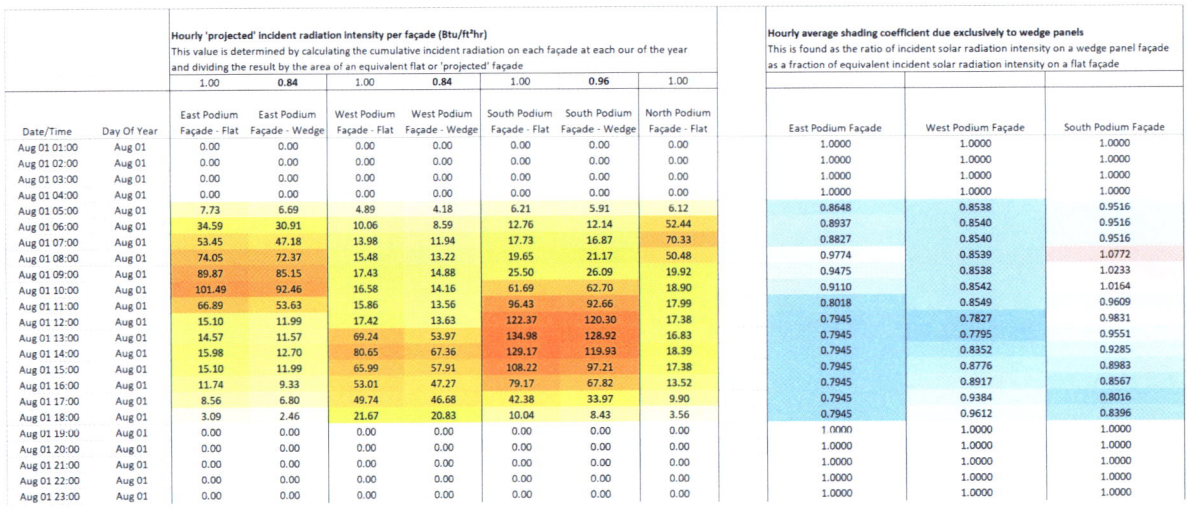

Cumulative Incident Radiation Intensity with Context Buildings — kWh/m² per year

Energy Modelling in Architecture

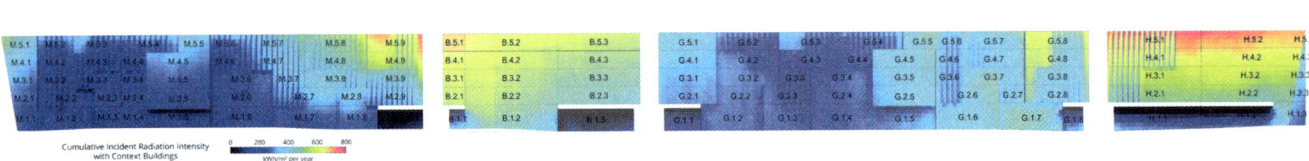

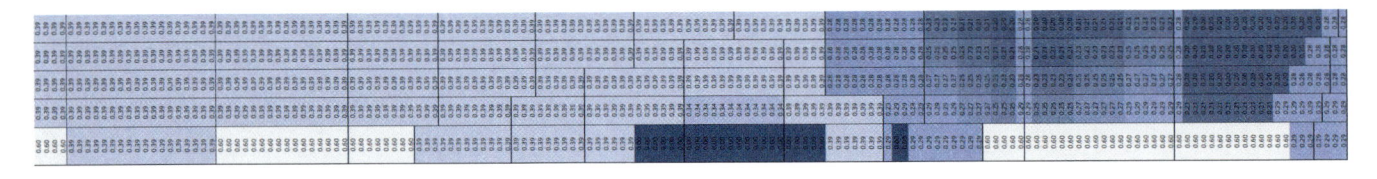

In pursuit of the ideal building energy modelling workflow

Both examples highlight how KieranTimberlake have synthesised traditional energy modelling methods in novel or atypical ways to better quantify the energy contribution of typically neglected design features. At a broader scale, they also illustrate shared methods that begin to reframe and resolve the persistent technical and social challenges that occur in energy modelling workflows. Thus, over the course of these studies, KieranTimberlake began to ask what an ideal workflow between the design and analytical disciplines would look like.

How much further does the team need to go to understand the gaps between their respective domain expertise? What are the limitations of the traditional workflow used by conventional firms relative to contemporary design and energy modelling paradigms? Developing an ideal workflow that resolves these limitations is a critical step towards generalizing and integrating bespoke analysis methods into traditional energy modelling practices.

A traditional energy modelling workflow generally consists of the following steps:
1. Performance and goal-setting is established, either according to energy standards in building codes, or imposed by the client or proactively by the architect.
2. The energy modelling consultant translates the architectural model information into a discipline-specific analytical model for the purpose of energy, and energy-related performance, simulation.
3. The simulation results are then communicated back to the architecture team in the form of images or reports consisting of high-level summaries of tabular or numerical data that then forms the subsequent round of design options.

Figure 4:
Typical energy modelling workflow – linear

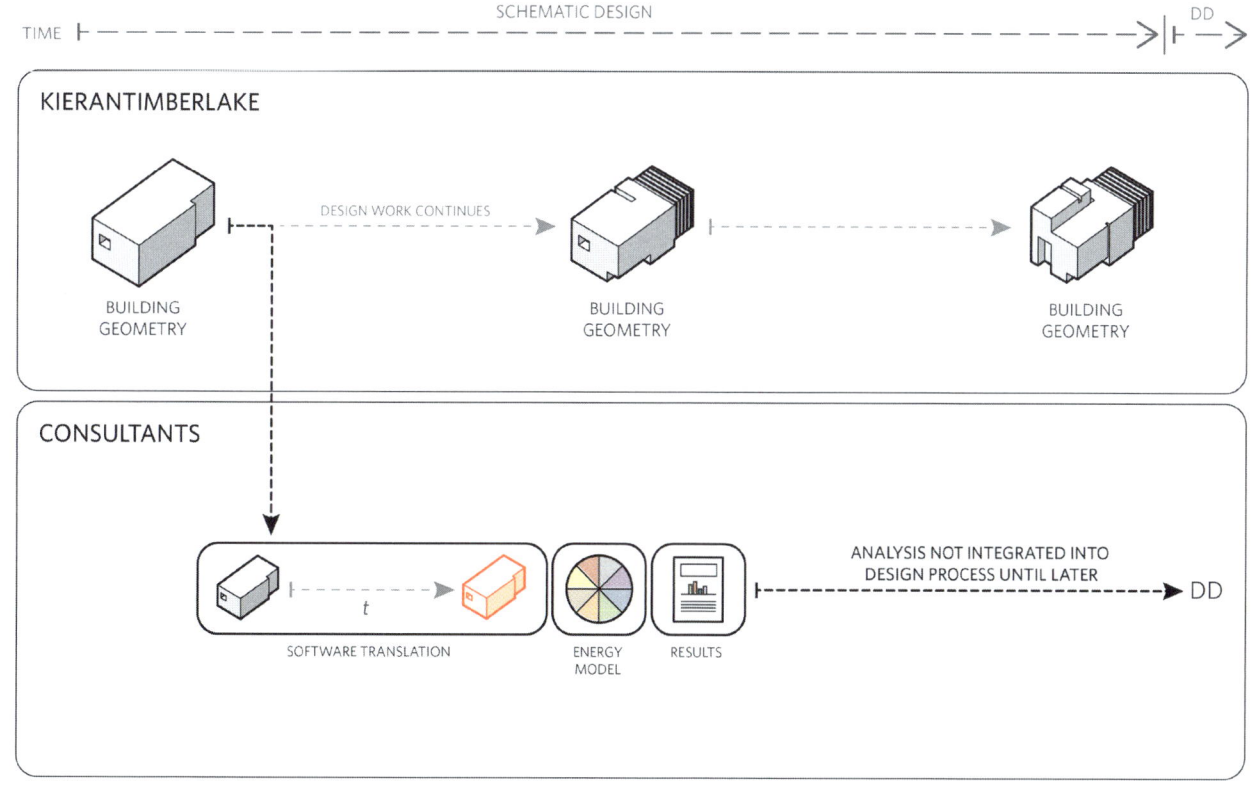

Within the broader context of state-of-the-art design paradigms, this energy workflow falls within what is often called the performative design paradigm, where an attempt is made to evaluate building form against established or emergent performance criteria rather than aesthetic considerations.[10] While firms that implement energy modelling integration theoretically champion such goals, in practice traditional workflows severely constrain the ability to drive the design process through iterative performance analysis. Flagler and Haymaker studied the limitations of traditional performative workflows based on a series of directed interviews with architects and engineers from a leading firm.[11] They found that during the typical conceptual design phase, only three design-analysis iterations are assessed, with each one generally taking a month to produce.[12] In their survey, they found that architects and engineers spend only 8% of their time interpreting results. The majority of their time is monopolised by building options, running analysis, and representing and documenting existing information.

Technical and social challenges that arise from energy modelling

The limitations of the traditional energy modelling workflow reflect intrinsic constraints and opportunities within the energy model itself, specifically the physics-based models traditionally used by the architecture, engineering and construction (AEC) industry. Physics-based models predict energy through a theory-driven approach by modelling the flow of heat and mass in the building through numerical methods. This is in opposition to the purely statistical, data-driven approach sometimes used in urban-scale energy modelling, where an assumed mathematical relationship between energy consumption and building features is applied to predict consumption.[13] From the perspective of design integration, the physics-based approach offers significant advantages over the statistical approach, as it can be built from more easily available design information and – in theory – allows the evaluation of specific design changes. There are however technical and social factors inherent in the way physics-based models are used that lead to conflicts in its application as a design tool.

Generally, technical conflicts arise from the difference between the architectural representation of the design and its analytical abstraction. The geometry composition of the typical design model reflects the prioritisation of construction logic and visual documentation. Digital model spaces are built up from individual component parts with an eye to representation in two and three dimensions. In contrast, the geometry logic of analytical models is dictated by the need to efficiently budget surface heat and mass transfer via numerical simulations. From this perspective, there is a twofold increase in the complexity overhead associated with detailed building geometry: time spent on the manual creation of the model, and increased time in its computational simulation. Thus, building geometries are heavily simplified in analytical models, which in turn constrains the evaluation of detailed or rapidly iterated design geometries.

The social constraints in this process arise from the fact that the composition of physics-based models varies significantly according to discipline-specific goals and heuristics of the energy modelling expert. These conceptual shifts cause confusion and conflict when the design team seeks to leverage the energy model for design studies.

KieranTimberlake consider this a social issue, because it pertains to how disciplinary knowledge from multiple stakeholders is translated into specific parameters and design solutions to address multiple, often competing, discipline-specific objectives.[14] For example, the energy modelling requirements to predict building operational energy vary significantly from those required to size mechanical systems. The former attempts to model the whole building as faithfully as possible, and simulates operations over the course of a year, while the latter might disregard realistic mechanical systems and simulate only an hour of extreme weather in the year. Furthermore, during the sizing process, the simplification or elimination of design features is further exaggerated relative to the engineer's tolerance for relying on non-mechanical factors to achieve prescriptive indoor comfort conditions. Without expertise in mechanical systems engineering, it is difficult for a designer to challenge, or design around, such constraints.

Therefore, physics-based energy models are well-suited to iteratively model and evaluate design feature changes but are limited by technical and social challenges, arising from gaps in the analysis and architecture domains. Specifically, technical challenges arise from the gap between the design and analytical model, which limit the ability to evaluate specific design features. Social challenges arise from the gap in discipline-specific knowledge and goals between designers and technical consultants, which obscures what can be evaluated and how that evaluation can be integrated in the design domain.

If the technical and social conflicts are eliminated, the project team can implement the ideal energy modelling workflow depicted in Figure 6. In contrast to the traditional workflow, the number of iterations and types of analysis overall have increased. Optimal design solutions can be found by traversing larger combinations of design and performance metrics. While the ideal workflow remains aspirational at KieranTimberlake, the practice has partially addressed the obstacles to its achievement by reducing the technical gap between the design and analytical model, and the social gap between disciplinary-specific knowledge and goals.

Specifically, the technical gap is being addressed through the development of a software application plugin for Revit that translates building information models (BIMs) into OpenStudio BEMs. This tool builds on previous work translating BIMs to eQuest energy models.[15] The resulting BEM is shared with external consultants in the form of a gbXML file, an interoperable building energy schema. This removes both the complexity of manually translating models and enables design teams to construct and critically interrogate the translation of architectural design features into analytical features. Furthermore, the shared model allows designers to run in-house performance studies, testing design features and incorporating them back into the shared model in a dynamic and nimble manner.

Resolution of the social conflicts requires multiple institutional and cultural changes to bridge the gap between discipline-specific knowledge bases and goals. Thus, at every design phase KieranTimberlake include a formal process map guiding energy modelling integration, expert assistance offered by an in-house building performance simulation (BPS) team; mandatory energy modelling meetings with external energy modellers; and regular in-house workshops on energy and energy-related performance topics to educate design teams. KieranTimberlake's strategies serve to educate designers and modellers, and build a shared vocabulary and culture around critically integrating energy performance into the design process. This social process is in turn closely tied to the technical innovations described above. Specifically, it relies on the novel synthesis of discipline-specific tools to act as a transdisciplinary platform for designers and modellers to share and question previously siloed concerns.

However, these gaps between technical model and discipline-specific knowledge have only been partially resolved. Limitations in the contemporary model exchange schemas prevent the sharing of the entire energy model, which enforces disciplinary separation of energy modelling enquiries. Specifically, KieranTimberlake's in-house energy studies tend to focus on demand-side energy modelling due to their inability to model detailed HVAC systems. For equipment-heavy projects like laboratories or hospitals, this continuing gap stifles energy-centric design strategies by concealing the building element responsible for most of the energy consumption from the design team.

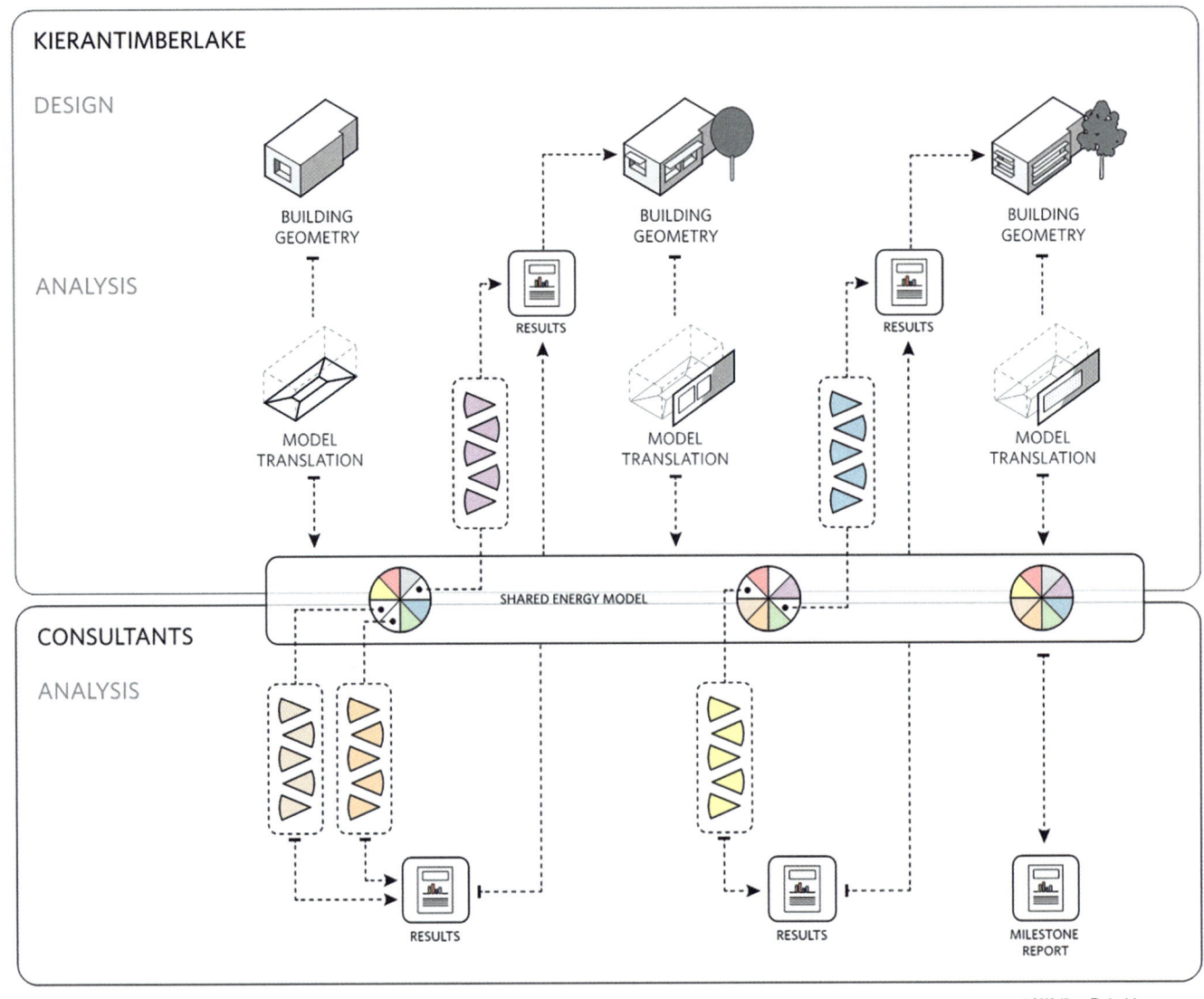

Therefore, while the advances in technical and social strategies have significantly reduced the time and ambiguity governing energy model and knowledge hand-offs, persistent obstacles remain. However, the methods outlined are not the only strategies being explored to improve the way designers test the performance impact of design features. KieranTimberlake have observed several emergent trends in the AEC industry that run parallel to this overarching effort:

- The integration of increasingly sophisticated data science methods, in particular machine learning, in design and performance modelling contexts. Such methods can improve the speed and ease of running simulations, as well as abstracting and extracting performance strategies from the resulting data.

Figure 5:
Ideal energy modelling workflow – iterative

Energy Modelling in Architecture

- The use of cheap, high-quality sensors to collect post-occupancy data in new and existing buildings. Sensor data can help narrow the gap between actual and predicted performance, a persistent problem that occurs due to intrinsic uncertainties in weather patterns, occupant behaviour and other processes.[16] This data can be used to create more robust inputs for simulation, provide contextual data to better represent uncertainty and variance in modelled scenarios, and remove obstacles to the use of statistical, data-driven energy models.
- The developing culture of shared or co-developed tools and workflows that resolve the technical fragmentation of building simulation tools. Culture here refers to at least two trends in particular: the increasing technical capability of AEC workers, and the commercial success of software released with permissive licensing. The continuing growth of both have democratised tool-building, and in turn flattened the hierarchical, siloed nature of domain-specific tools and their application.

While the practice continues to struggle to achieve a seamless and efficient feedback loop between analysis and design today, the continuing growth of these AEC trends implies there are multiple paths to its resolution. Central to each path is the building energy model itself, not only as an analytical tool, but as a design tool that, properly deployed, is one of the most effective strategies to reduce carbon-based building energy use. The dissemination and optimisation of building energy modelling workflows and tools is an important challenge to tackle through research and practice.

Towards a new shared workflow culture

KieranTimberlake's discussion of an evolving socio-technical approach that takes into account the importance of considering urban landscape contextual uncertainty and variability, often disregarded by modellers, begins to highlight the need for architects' early involvement and engagement in the energy modelling process. 'Building a shared vocabulary' is seen as a potential route to a truly integrated process to understanding and examining a building's performance through shared analytical and design workflows.

Continuing the discussion on the importance of shared analytical and design tasks across workflows, the following section by Henning Larsen considers the skills, knowledge and resources needed to enable an integrated approach to energy analysis and modelling at early stages of design. An early recognition, that multiple disciplines, including engineering, IT and sustainability consultancy were needed, enabled the development of Henning Larsen's approach to optimising building performance.

KieranTimberlake Solutions to Technical + Social Challenges in the Energy Model Process

Figure 6:
The energy modelling workflow

ENERGY MODEL CONTRACT

Establish
Ensure modelling contract responds to design questions and timeline.
What types of analysis will be performed for each phase?

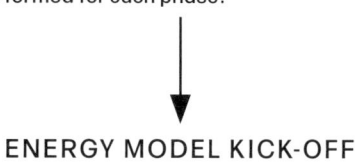

ENERGY MODEL KICK-OFF

Establish a close working relationship
Early design phase meeting to define energy goals, system boundaries, model input definition, energy conservation measures (ECMs), modelling software and format, process for model translation, schedule and reporting.

Identify the question
What is the purpose of the model? Does it make sense to perform a whole building energy model at this stage or a targetted analysis on a key parameter? Do the metrics answer the design questions?

Shared inputs
Which inputs can be improved beyond the default model assumptions?
Who is best equipped to provide these inputs?

ANALYSIS

Interoperability
Can the architecture model be easily translated into the beginnings of an energy model to reduce time spent on geometry construction on the energy modelers behalf?
Is there a way to ensure that the energy modeler is always referencing the latest architectural model?

REVIEW

Quality assurance
Review inputs used in the model and discuss preliminary results. Discussion of results before the report is written will ensure better interpretation for larger design team. Allow the energy modeller to respond to comments and improve delivery of final energy model.

DELIVERY

Reporting + interpretation
How is the data represented? Does it directly inform design? Can the data be spatialised?
What metrics are used? Does the report provide all of the outputs required for benchmarking or for other firm interests?

Chapter 6: Henning Larsen – discovery and experimentation

Over the past decade Danish standards for energy usage have become increasingly ambitious, establishing tighter requirements for a building's performance. To meet these standards and ensure a comfortable, healthy indoor climate, mechanical climate control systems alone are not sufficient. Henning Larsen's focus on sustainability has broadened over the years, and they now evaluate how a building, as the sum of its materials, design and fundamental volumes, can create a comfortable indoor environment at the lowest cost.

Figure 1:
Danish energy performance framework (kWh/m²/yr)

The history of Danish energy performance framework (kWh/m²/yr)

	2008	2010	2015	2020
150 m² single family home	84.7	63.5	36.7	20
1000 m² residential block	72.2	54.2	31	20
10,000 m² public school	95.2	71.5	41.1	25
1000 m² office building	97.2	73.0	42	25
300 m² daycare centre		76.8	44.3	25

Table. Energy requirements for dwellings, offices, schools, institutions, etc. This table shows the maximum allowed demand of the building for energy supply for heating, ventilation, cooling and domestic hot water and lighting per m² of heated floor area, as measured over the course of the year. It illustrates Danish national standards' increasingly demanding goals for energy performance.

Initially, Henning Larsen's approach to energy modelling focused primarily on maximising the thermal retention properties of individual building components, based on the idea that the majority of a building's energy usage came from compensating for thermal losses. As increasingly advanced insulation technology made heat retention easier, the practice began to focus more on optimising the energy-intensive systems needed to maintain indoor comfort. The practice also started to recognise the need to consider energy performance at earlier design stages.

New knowledge in energy modelling
As attitudes towards energy modelling and optimisation in the industry shifted, the practice recognised the need for a response that not only met the new national requirements, but also created a framework to guide their future growth. To expand the knowledge base informing project designs, Henning Larsen began an industrial PhD programme, in which they co-hosted PhD students with backgrounds in engineering, IT and sustainability to help synthesise and integrate energy-efficient building and design methods into the overall design process.

When the first three PhD students arrived at Henning Larsen in 2008, they were the first specialised engineers to work at the firm in its 50 years of existence. Incorporating unconventional knowledge and skills into standard architectural practice established Henning Larsen's approach to energy modelling. The proactive search for outside knowledge is a defining element of Henning Larsen's approach to energy optimisation, and a vital component for creating long-lasting, sustainable architecture. Aesthetics is quantified through energy reduction, and energy reduction is qualified through aesthetics. Combining the theoretical knowledge of the PhD specialists with the practical approach of the architect has been found by the practice to facilitate a more enduring architecture.

Theory and praxis – a contemporary approach to energy optimisation

The methodology used by Henning Larsen is that of integrated energy design, which considers energy consumption as a priority from the earliest design stages. Findings from the practice's doctoral research projects suggest that 40 to 50% of a building's energy usage is determined within the architectural design concept. This means that energy considerations carry the greatest impact if they are considered as early as possible in the design process. Integrated energy design is a process of constant evaluation, adjustment and reassessment of the relationship between a building's design and its energy performance. Energy optimisation studies are continually conducted by the in-house engineering team, as well as in close collaboration with external consultants such as BuroHappold. In this way, the architectural design concept is developed on the basis of actual data, analysis and simulations, which provide an aspect of quality control for the building.

This approach to integrated energy design is best illustrated by a three-tiered pyramid, which showcases the energy priorities and their organisation in the design process.

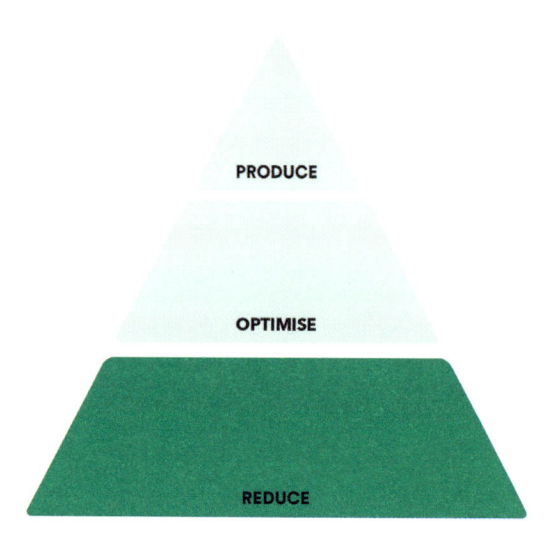

Figure 2: Integrated energy design pyramid

The pyramid structure illustrates how the different agents are divided into reducing, optimising and producing measures. As the base, reducing energy usage through fundamental building aspects represents the first, and ultimately most influential, area of design refinement.

The most significant energy reductions emerge through passive strategies, such as optimising building orientation, geometry and natural airflow. As fundamental properties of the building, such adjustments remain effective throughout the building's life, but they require thorough preparation and an intelligent use of resources. Therefore, reducing energy usage through these passive means is the first logical step, and the foundation of the pyramid.

Figure 3:
Reduction – four forms

REDUCE:

Context
Wind, water, daylight, noise and pollution.

Shape
Geometry, orientation zones and daylight.

Structure
Space, daylight, main functions, zones and construction.

Facade
Daylight, technology, indoor climate and user behaviour.

Within the foundational Reduce level of the three-tier pyramid, the four subsets of Context, Shape, Structure and Facade lay out the most important elements of reduction when optimising a design's projected energy performance. These core areas encompass considerations that make the most significant difference in reducing a building's passive energy usage, and many of them represent decisions made in the very early stages of the design phase.

Adjusting energy performance through components' technical installations, the focus of the Optimise level of the pyramid, incurs additional costs. Investing in a high-performance ventilation system is necessarily more expensive. However, costs are recouped within a relatively short period in the form of lower operating costs and reduced CO_2 emissions. As such, the initial investment in higher-quality, better-performing HVAC systems and other building components is often a more sustainable choice, both in terms of energy usage and operational finances.

The top of the pyramid represents the integration of renewable energy systems into the building. Components such as solar panels have a positive effect on the energy balance, but also cost a great deal, and are not currently long-term investments. Measures of this kind only create energy-related value, and not necessarily improved comfort value.

The objective of the integrated energy design method is to minimise energy usage through an iterative process where specialised knowledge drives the creative process, while aesthetics and space are the means of achieving significant energy reductions. In this way, both aesthetic and technical parameters play lead roles in performance optimisation.

The Kolding Campus building for the University of Southern Denmark is a model of this three-tiered approach, as it was designed to suit the 2015 energy requirements of Danish building regulations.

Following an integrated energy design process throughout the development of the Kolding Campus project enabled the designers to optimise energy performance at every level. The practice began with a basic geometry where the performance is set by the minimum energy standards for insulation, heat recovery, lighting systems, etc.

This example made use of passive, energy-reducing solutions integrated into the building design (the Reduce step) and partly through energy-producing elements such as solar cells, groundwater cooling and heating (the Optimise step). The project serves as an example of a compact building geometry and a holistic daylight strategy – the common denominator between architecture and energy – and has resulted in a notably efficient overall energy performance level.

Reference: 95 kWh/m²/year

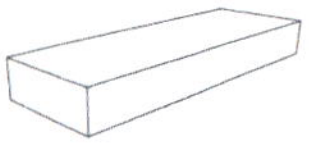

The project is based on the standards for traditional building as provided for in Danish building regulations BR08. This corresponds to an energy consumption of 95 kWh/m²/year. The objective for the project is to meet the 2015 energy requirements of Danish building regulations, corresponding to approx. 42 kWh/m²/year.

Reduce: 95 kWh/m²/year → 88.8 kWh/m²/year

Context
The new campus building in the centre of Kolding is located by the river and close to the harbour. Situated adjacent to Kolding Design School and International Business College Kolding, the new Kolding Campus will become part of a dynamic study environment.

Orientation and position
Kolding Campus is located in the north-east corner of the site. The rotated position of the building creates a sunny central plaza between the campus and the river and prevents a direct north-facing facade with no sunlight.

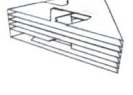

Geometry
The triangular shape of the building ensures an optimal use of square metres. The large rotated atrium provides the building with both ample daylight and a view to all world corners. At the same time, the atrium provides supplementary natural ventilation and night cooling.

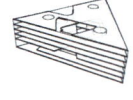

Daylight
Achieving the right amount of daylight in a building is a balancing act between use of large, open windows and shielded windows the orientation and design of the skylight protect the interior from direct sunlight – as too much light can have a negative impact in the form of increased cooling and ventilation requirements. The atrium provides optimal daylight conditions in the heart of the building.

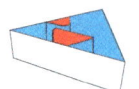

Functional layout
Kolding Campus offers a good, differentiated learning environment with different indoor air quality zones. The building has two climate zones. Teaching and administration facilities are situated in the zone closest to the facade, which has a stable indoor air quality. The atrium has a more fluctuating indoor air quality – allowing the users to sense the changing seasons.

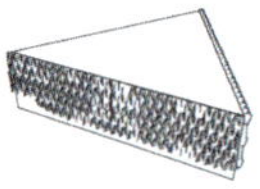

Facade design
As a part of the daylight strategy, a dynamic, mobile solar protection system has been developed for the facade. It consists of a light structure of movable, triangular elements, which regulate the daylight intake, as well as a heavier, well-insulated structure. The opening angle of the facade is approx. 50%.

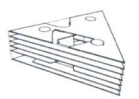

Heavy structures
Kolding Campus is part of a development project which examines how the thermal properties of concrete can be increased – and the energy consumption for heating and cooling thus reduced. In order to make optimal use of the thermal properties of concrete, the slabs are exposed where possible. This prevents large fluctuations in temperature and improves the indoor air quality.

Optimise: 88.8 kWh/m²/year → 57.9 kWh/m²/year

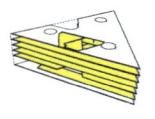

Lighting
Kolding Campus features needs-based lighting. Energy-efficient LED lighting has been applied in the entire building.

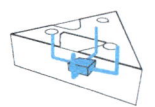

Mechanical ventilation
A mechanical, needs-based VAV (Variable Air Volume) ventilation system with high efficiency has been installed in the building. The system works together with the thermo-active structures. Vapour-permeable ceilings ensure a low pressure loss and reduse the amount of pipes and fittings.

Produce: 57.9 kWh/m²/year → 38.4 kWh/m²/year

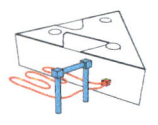

Aquifer Thermal Energy Storage (ATES)
Kolding Campus features a combined heating and cooling pumping system, which uses groundwater to regulate the building temperature. The fully integrated system works together with the other building installations, which for instance apply the outside air for cooling.

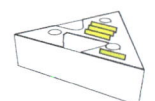

Solar cell system
A solar cell system on the roof produces electricity.

Figure 4:
Use of the optimisation levels in the USD Kolding Campus project

Energy Modelling in Architecture

Rhino
Computer-aided design software

Grasshopper
Parametric modeling for Rhino

Ladybug
Environmental plugin for Grasshopper

Honeybee
Connector to validated simulation engines

Energy +
Thermal modelling and building energy simulation

Radiance
Daylight simulation

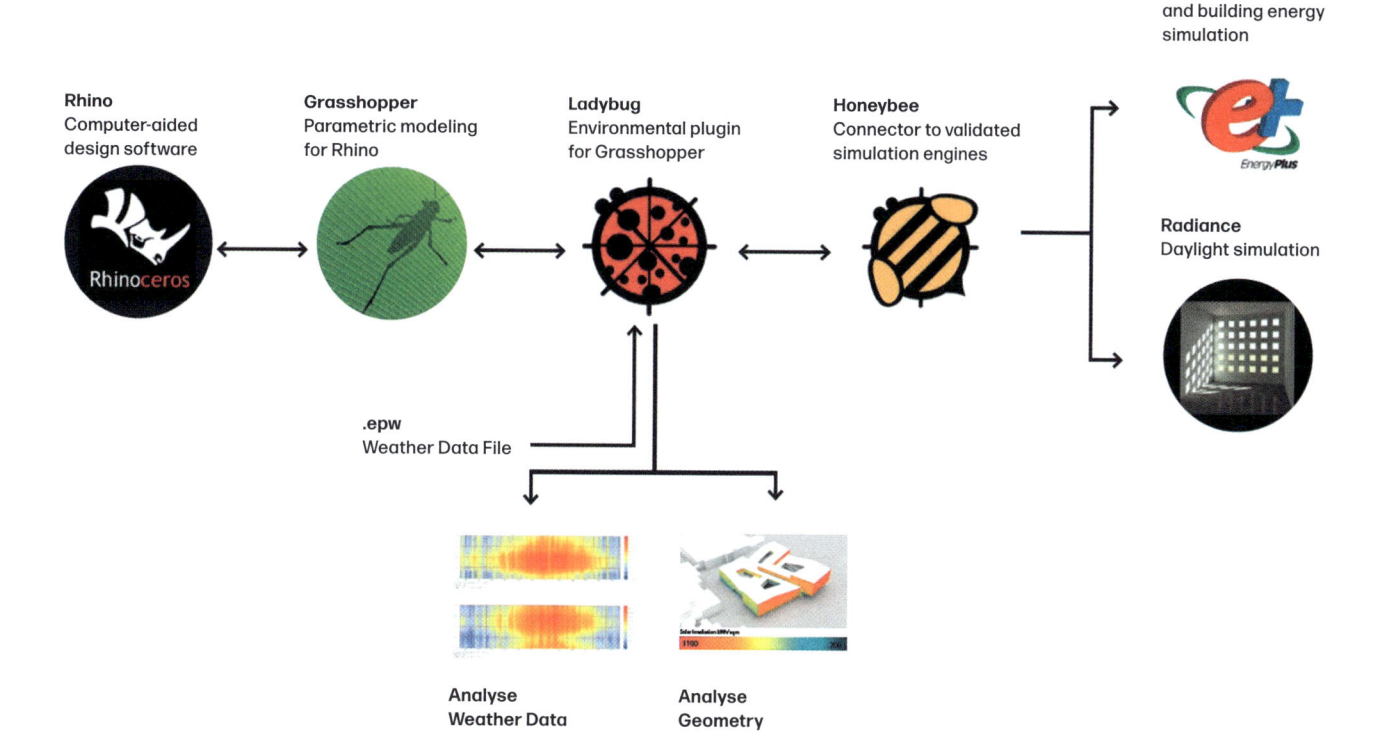

.epw
Weather Data File

Analyse Weather Data

Analyse Geometry

Figure 5:
Relationship between the use of software and workflows

Praxis and workflow

The day-to-day practice in integrated energy design relies on an open-platform technological approach that allows energy optimisation efforts to cross design phases, specialists and software. The relationship between the software implicated in Henning Larsen's workflow, and their respective functionalities, has been illustrated in Figure 5.

Allowing energy modelling to be a more integrated part of Henning Larsen's process required the practice to adopt a flexible workflow that could inform designs more quickly and intuitively.

While the process is rarely the same on different projects, a typical workflow is as follows:

1. Gather project information. The sustainability consultant gathers an overview of local code requirements and regional climate data. This is visualised through diagrams for architects and colleagues.
2. Together with architects, engineers and specialists, find a strategic sustainability concept that will be interwoven into the overall storytelling angle of the project. Clients want good publicity around their building, and strong storytelling helps consultants convince clients to prioritise sustainability more highly.
3. Perform quick analysis on architect's volume studies: Visualising how the different volumes perform in the early phases gives the architect an idea of how to further develop the concept. This visualisation enhances the architects' sketching process and makes proposals easier to transfer to energy consultants later on.

4. Facade studies on shoebox models: When finding a facade concept, early sketches are quickly assessed for daylight and energy performance. Risk of overheating, cold draughts and solar glare are valuable matrices to evaluate facade concepts. Here, several options are tested, evaluating HVAC dimensioning, and upfront costs and operational costs are compared. Typical questions asked at this stage are whether a smaller window is a better long-term investment, or whether chill beams can increase ceiling height and improve daylight.
5. Perform early-stage full-floor analysis: Analysis of selected typical floors to see compliance with codes and certifications helps with setting guidelines for remaining floors.
6. Further analysis: Shading types, eventual roller blind fabrics and glass types are selected based on energy models and daylight simulations. Facade details are assessed in THERM and HEAT2/3.

Carl H. Lindner College of Business, University of Cincinnati

Henning Larsen's recent work on the Carl H. Lindner College of Business at the University of Cincinnati in Ohio provides an example of the practice's approach to integrated energy modelling. Henning Larsen won the competition for the 22,500m^2 academic space in 2016 and began research on project site conditions in the same year. Opened in early autumn 2019, the building incorporates classroom spaces, private faculty offices and open social spaces for the wider campus community.

BuroHappold provided engineering expertise and consultancy on the project. The following section reflects their approach and input into the energy modelling process.

An engineer's perspective – BuroHappold Engineering

The design of the University of Cincinnati Linder College of Business utilised a variety of building performance modelling tools to inform the design process:

- Energy modelling – used to study the annual energy consumption created by the interaction of buildings systems, occupants and the campus central plant.
- Dynamic thermal modelling – used early in the design process to create quick studies of the building massing and spaces, with a focus on thermal comfort and thermal loading.
- Daylight modelling – to study the distribution of light through spaces such as offices, atriums and courtyards.
- Computational fluid dynamics – to study the movement of air and heat inside the building. This tool was also used to study wind around the building and the surrounding site context.

Each of these tools was deployed at various design phases of the project to provide the design team and client with information on the potential impact of developing particular design approaches.

We can reduce annual solar gain cooling loads by approx.

70 %

by locating the closed office towards north

OFFICE AREAS ARE PLACED NORTH
to avoid overheating, reduce cooling demand and utilise stable daylight.
- offices have high internal gains and thus high cooling demand
- by placing them north additional solar gains are being avoided
- lighting levels are optimum so lighting demand is also lower further reducing heat gains from lamps

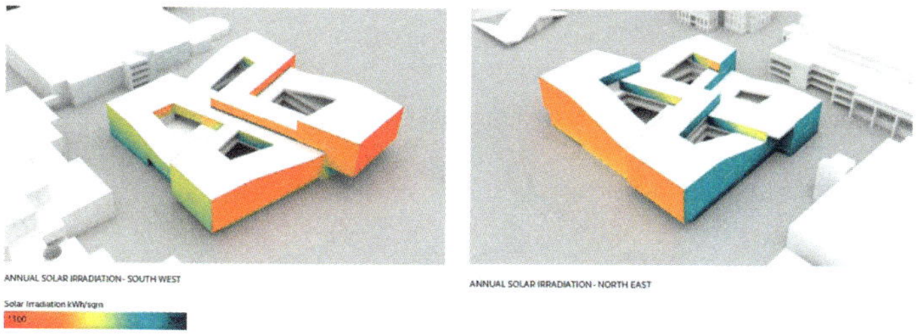

ANNUAL SOLAR IRRADIATION- SOUTH WEST

ANNUAL SOLAR IRRADIATION- NORTH EAST

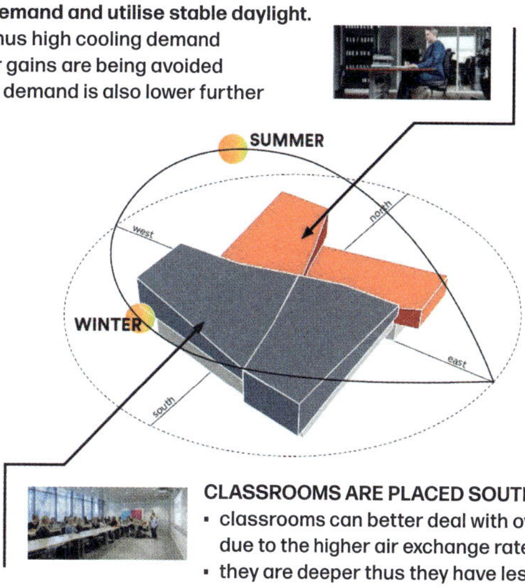

CLASSROOMS ARE PLACED SOUTH
- classrooms can better deal with overheating due to the higher air exchange rate
- they are deeper thus they have less solar gains/energy per floor area
- % of area with high sun exposure is small due to the rooms' depth/size and can be controlled with shading devices

Figure 6:
Location of classroom and office areas in Kolding Campus

During the early conceptual design stage, a wide range of massing strategies were proposed. These ranged in appearance and complexity, each looking to balance the internal programmatic needs with desired performance characteristics. During these early stages the structural and mechanical engineer contributed to the design dialogue alongside the architect, offering advice on the massing strategies.

Using the above performance criteria and supported by conceptual-level daylight and dynamic thermal modelling, a massing strategy was selected to be advanced from concept to schematic design. With a basic massing strategy in place, the design team began to evaluate several facade design approaches. These facade designs were evaluated in conjunction with mechanical system options.

Energy targets

The University of Cincinnati had a series of energy performance targets, including the pursuit of a USGBC LEED version 4.0 rating. As part of a holistic strategy to achieve this goal, a series of energy targets were discussed during the design process. These discussions started as building energy-use intensity (EUI) targets measured as KBtu/sf/yr. As the schematic design was being progressed, specific energy conservation measures were proposed and studied using dynamic thermal modelling, energy modelling, daylight modelling and computational fluid dynamic modelling.

Some of the energy conservation measures included:

- radiant strategies for thermal control
- window-to-wall ratio tuning to balance solar and thermal loads with low-energy HVAC strategies

- glass and spandrel tuning to define SHGC (G-factor) and U-value
- daylight control
- dedicated outdoor air
- integration with campus central plant versus building-level heating and cooling plant equipment.

Thermal comfort, daylight and energy

During the schematic design stage, a dialogue between the mechanical engineering team and the architecture team revolved around the relationship between the use of a radiant-based HVAC strategy and the ability of the facade to control solar gain at the perimeter. The massing of the building had evolved to provide ample access to daylight for individual faculty offices, conference rooms and classrooms. Access to daylight was balanced with perimeter thermal control strategies and overall building EUI targets.

The use of radiant systems to provide sensible heating and cooling as well as thermal control combined with a dedicated outdoor air system is a low-energy HVAC strategy. An active chilled beam is an example of this type of HVAC strategy. This uses the natural buoyancy of air in conjunction with a ducted supply of conditioned outdoor air to provide both sensible heating and cooling to a space as well as latent control. Using the natural buoyancy of air across a chilled beam's surfaces reduces fan energy. To maximise the efficiency of this strategy it is important to understand the thermal capacity of an active chilled beam. This capacity is limited by the rate that the beam can induce air across the heating/cooling surfaces using buoyancy. This induction rate is a critical factor in determining how much thermal load per square metre an active chilled beam strategy can

Figure 7:
Sensitivity analysis of whole-building glazing

Window-to-Wall Ratio

A sensitivity analysis of whole-building glazing percentages for the project's building facades in relation to heating and cooling load was completed through iterative parametric building simulation runs (40%, 50%, 60% and 70% WWR at each orientation). As expected as the WWR increases beyond the prescriptive energy code maximum on the building (40%) we observe increasing levels of peak load intensity and duration.

As shown in the figure above, internal gains are observed to be the dominant whole-building geometry and varying WWR (ranges from 52% to 44% of the peak building cooling load respective to 40% WWR and 70% WWR). Second to internal gains, solar gains increasingly vary from 25% to 33% of the peak building cooling load respective to 40% WWR and 70% WWR.

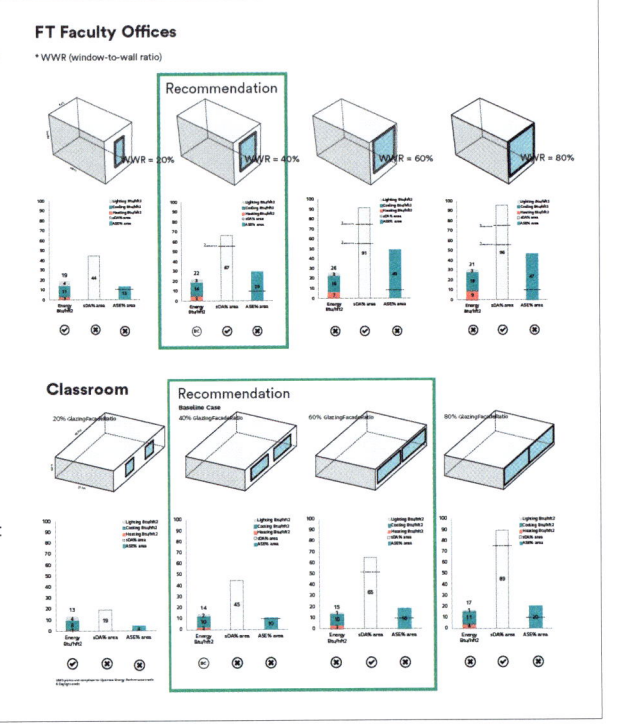

58   Energy Modelling in Architecture

manage. The mechanical engineering team proposed an active chilled beam strategy to condition many zones of the building, such as faculty offices. Most of the faculty offices were located around the perimeter of the building to support access to views and daylight.

The determining factor in evaluating an active chilled beam's energy-saving potential is defined by the solar gain to the thermal control zone. If the solar gain entering a thermal control zone through facade glazing exceeds the thermal capacity of an active chilled beam it is necessary to increase the ducted airflow to the beam. This increase results in the consumption of more fan energy, larger ducts and larger air-handling units. It was a goal of the design team to balance the solar gain with the thermal capacity of an active chilled beam strategy at perimeter zones.

Performance modelling
A series of dynamic thermal models were developed to study balancing the thermal capacity of active chilled beams with solar gain entering perimeter thermal control zones. These models allowed the design team to study the facade window-to-wall ratio, glass selection and HVAC strategies with the goal of maintaining thermal comfort while minimising HVAC energy. Rather than attempt to build a whole-building energy model during schematic design, a series of simplified dynamic thermal models were constructed. These models focused on a few space types at the perimeter, such as faculty offices. IES VE Apache was used for this project. The dynamic thermal model was linked with the IES VE Apache HVAC model to calculate the active chilled beam induction rate and ducted fresh air supply to the spaces. Finally, these spaces were also modelled in a daylight simulation software. Collectively, these tools allowed the design team to study the relationship between:

- window-to-wall ratios ranging from 30 to 50%
- SHGC coating on the glazing units' management of solar gain in a manner that optimises the energy-saving potential of active chilled beams
- the resulting duct size from the dedicated outdoor area and resulting floor-to-floor height
- the potential thermal comfort of an occupant in the various spaces/thermal zones
- the resulting energy-saving potential and cost savings.

While the dynamic thermal models provide valuable information about the annual energy consumption and comfort conditions in the thermal zones studied, there are limitations in the detail that these tools can capture. To further study the relationship between the facade and the HVAC strategy it was important to consider the occupants' experience, which was captured through thermal comfort studies. Thermal comfort of occupants is influenced by convection, conduction and radiation. A computational fluid dynamics (CFD) study of an individual faculty office was created. A CFD model is a form of energy modelling that captures the movement of heat through air with much more detail, but represents a single instance in time. In addition to convection, CFD can simultaneously capture the transfer of heat through conduction, such as through a facade, and radiation from the sun or building internal surfaces. This modelling tool can provide detailed insight to occupant thermal comfort by simulating airflow patterns, velocity and temperature. The location of the active chilled beams, lights and return grilles in the ceiling plane was studied to maximise comfort and minimise the number and length of active chilled beams.

Figure 8:
Thermal zone modelling study

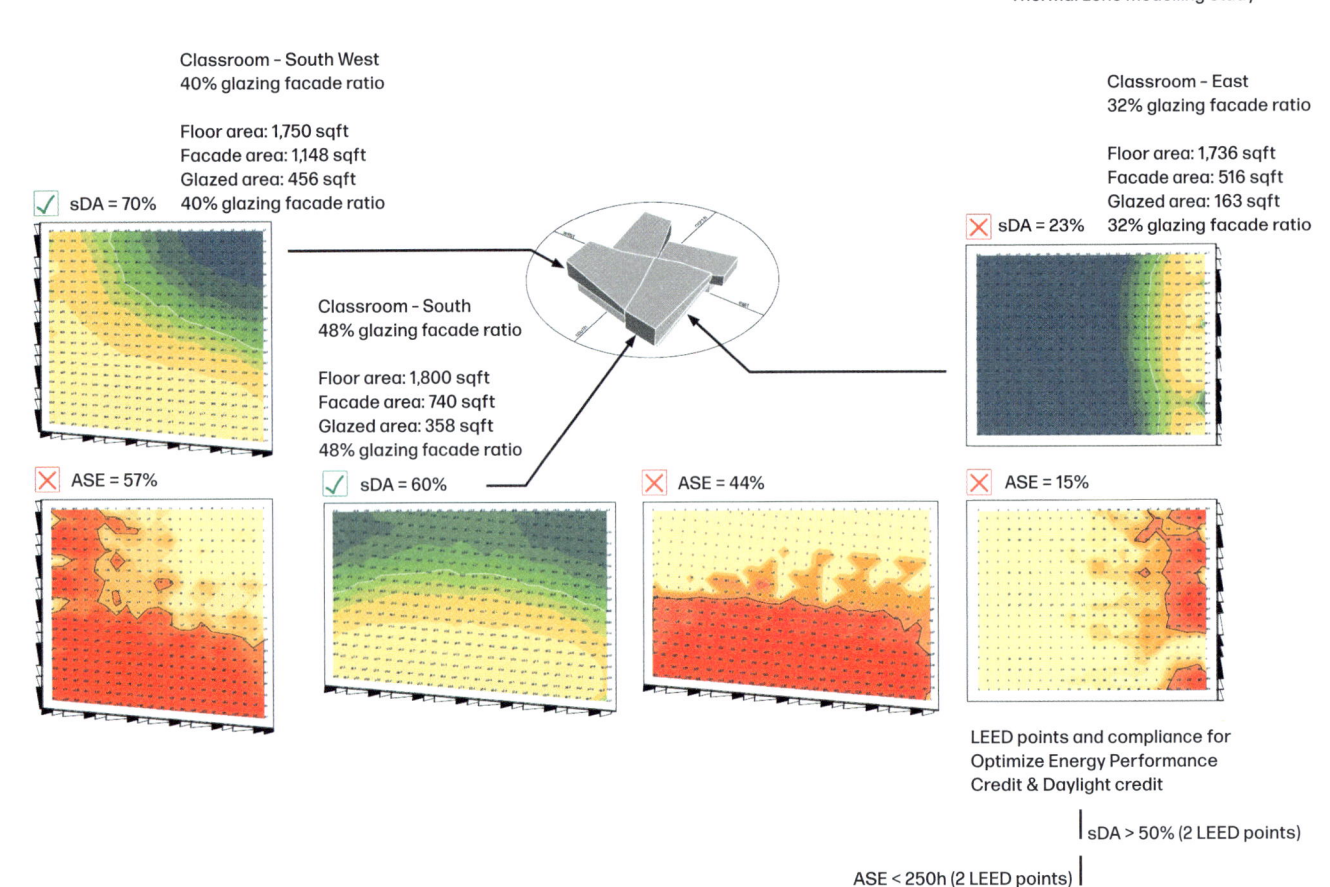

Classroom – South West
40% glazing facade ratio

Floor area: 1,750 sqft
Facade area: 1,148 sqft
Glazed area: 456 sqft
40% glazing facade ratio

✓ sDA = 70%

Classroom – East
32% glazing facade ratio

Floor area: 1,736 sqft
Facade area: 516 sqft
Glazed area: 163 sqft
32% glazing facade ratio

✗ sDA = 23%

Classroom – South
48% glazing facade ratio

Floor area: 1,800 sqft
Facade area: 740 sqft
Glazed area: 358 sqft
48% glazing facade ratio

✗ ASE = 57%   ✓ sDA = 60%   ✗ ASE = 44%   ✗ ASE = 15%

LEED points and compliance for
Optimize Energy Performance
Credit & Daylight credit

| sDA > 50% (2 LEED points)
ASE < 250h (2 LEED points) |

Confidence to build

The collective use of dynamic thermal modelling, daylight modelling and computational fluid dynamics modelling early in the design process allowed the design team to propose a low-energy HVAC strategy aligned with the facade performance. The use of iterative modelling enabled the design team to tune the building facade to maximise the energy-saving potential of a low-energy radiant strategy represented by the active chilled beams. This was done while maintaining access to daylight and occupant thermal comfort. The performance modelling tools informed the client of the impact and outcome of the various design decisions in a holistic way, balancing the design issues of accessibility, views, access to daylight, acoustics, facade areas, access to campus utilities and infrastructure, time of use patterns, security and thermal comfort.

The resulting outcome of combined efforts between disciplines meant that the north-oriented open offices were calculated to save $10,000 in operational costs each year, consuming around 45% less energy per person compared with conventional office spaces. Overall, the design predicted a likely reduction of the building's thermal heating loads by up to 70% compared to fully sealed office spaces.

Figure 9:
Adding in detail

Guidelines, Window-to-Wall Ratio

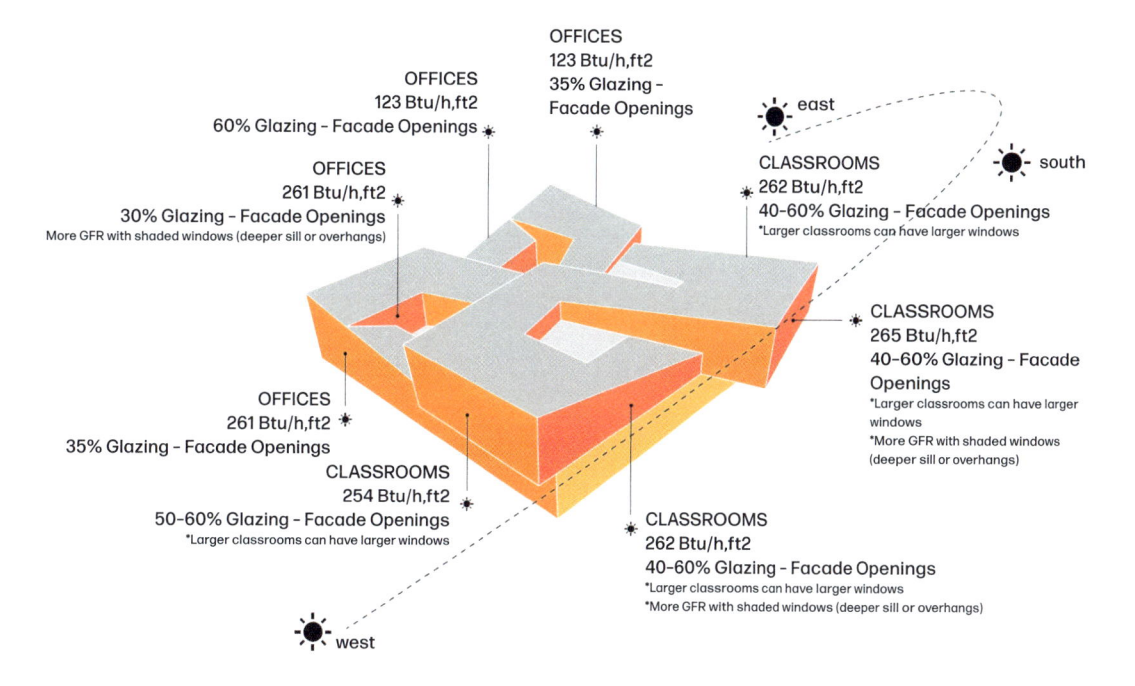

OFFICES
123 Btu/h,ft2
60% Glazing - Facade Openings

OFFICES
123 Btu/h,ft2
35% Glazing - Facade Openings

east

OFFICES
261 Btu/h,ft2
30% Glazing - Facade Openings
More GFR with shaded windows (deeper sill or overhangs)

CLASSROOMS
262 Btu/h,ft2
40-60% Glazing - Facade Openings
*Larger classrooms can have larger windows

south

CLASSROOMS
265 Btu/h,ft2
40-60% Glazing - Facade Openings
*Larger classrooms can have larger windows
*More GFR with shaded windows (deeper sill or overhangs)

OFFICES
261 Btu/h,ft2
35% Glazing - Facade Openings

CLASSROOMS
254 Btu/h,ft2
50-60% Glazing - Facade Openings
*Larger classrooms can have larger windows

CLASSROOMS
262 Btu/h,ft2
40-60% Glazing - Facade Openings
*Larger classrooms can have larger windows
*More GFR with shaded windows (deeper sill or overhangs)

west

Figure 10:
Illustration of daylight analysis of site

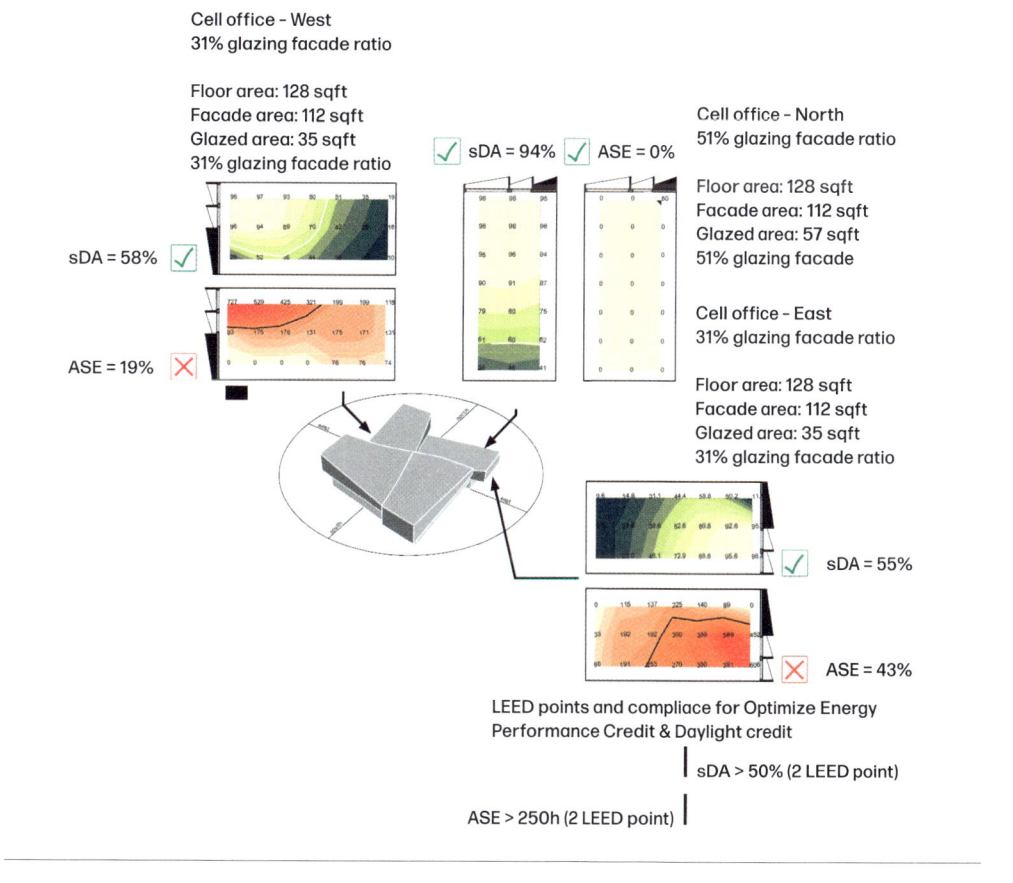

Cell office - West
31% glazing facade ratio

Floor area: 128 sqft
Facade area: 112 sqft
Glazed area: 35 sqft
31% glazing facade ratio

sDA = 94% ✓ ASE = 0% ✓

Cell office - North
51% glazing facade ratio

Floor area: 128 sqft
Facade area: 112 sqft
Glazed area: 57 sqft
51% glazing facade

sDA = 58% ✓

ASE = 19% ✗

Cell office - East
31% glazing facade ratio

Floor area: 128 sqft
Facade area: 112 sqft
Glazed area: 35 sqft
31% glazing facade ratio

sDA = 55% ✓

ASE = 43% ✗

LEED points and compliace for Optimize Energy Performance Credit & Daylight credit

sDA > 50% (2 LEED point)

ASE > 250h (2 LEED point)

Working across and between disciplinary boundaries

The new Carl H. Lindner College of Business, as described above, enabled Henning Larsen and BuroHappold to develop an integrated collaborative approach to sustainability through a focus on human comfort, demonstrating how energy modelling exists at the intersection of technological innovation, real-world research and social consideration.

Approaches to energy optimisation, both in concept and practice, have developed in Henning Larsen's practice dramatically in recent years. Sustainable, enduring architecture is viewed by the practice to reflect its place in a larger equation of environmental, social and contextual considerations. This broader perspective acknowledges the social responsibility of architecture – calling for a greater consideration of how the spaces designers create influence those who use them. Ultimately, the focus should be on the social effects of designs.

The above discussion illustrates Henning Larsen's point that buildings are as good as designers allow users to make them. As designers shape the building, so does the building shape its occupants – therefore, energy optimisation must include human considerations, reflecting an understanding of how local users engage with the built environment. Architecture can foster inclusivity and new connections, providing the physical structure for social and environmental good.

The following section by Architype further extends this approach to collaborative, integrated design rooted in a highly contextual, socially aware and climatically responsive approach to energy modelling.

Chapter 7: Architype – a question of culture

Over the past three decades, low-energy, low-carbon design has become mainstream, driven not just by the current political agenda and climate emergency, but also by clients with buildings that rarely perform as predicted.

Accurate energy modelling is critical to creating a successful building. However, buildings that overheat or waste energy are still far too common. In the past ten years Architype have been using the Passive House energy and comfort standard as a basis to deliver buildings that perform well. Passive House design and construction means people get a building that is built well, is unlikely to overheat or develop draughts or mould, and has plenty of fresh air and daylight: in essence, a healthy building that uses the bare minimum of energy and carbon. Designing to Passive House Standard requires a change in mindset by architects and the design team. It means understanding the rules of building physics and the importance of form and orientation, and designing to a much higher specification of airtightness and thermal performance. Critically, it also means a joined-up strategy for ventilation using heat recovery. This isn't necessarily a constraint, but rather can be a way of liberating designers to have a high degree of confidence in the moves they are making for a particular site. Unlike a naive design, which can entail a frustrating and often reductionist process of making a scheme comply to technical performance standards, the use of Passive House and Passive House Planning Package (PHPP) modelling, combined with a good knowledge of construction, enables creative optimisation of design.

The design vision and the building's energy efficiency need to be aligned from the very beginning. Design ideas that start with pencil and paper are rigorously tested to check energy efficiency before they make the final cut. Trying to make a building achieve the Passive House Standard retrospectively is very difficult if the design principles are not right to start with.

Increasing in-house skills is important. Architype work very closely with engineers as part of the design team, but also have a good technical understanding of how to make buildings work. They now have ten certified Passive House designers, and they also employ dual-qualified architects and engineers and invest in training for staff in tools such as BIM and WUFI, which assesses hygrothermal performance and thermal bridging. Thermal bridge modelling can calculate the temperature distributions and heat flows in a building structure and mitigate the risk of condensation or heat loss. Aspects such as daylight and ventilation modelling, and embodied carbon analysis, can prove particularly useful in developing an informed approach to building design right from the earliest stages. The practice employ a technical manager and research affiliate whose role is to build knowledge across the practice and offer insight to teams as designs develop.

PHPP is one of Architype's key energy modelling tools. This optimises the energy efficiency of each design, along with BIM and software such as Revit to provide a holistic approach and give an insight into how well the building will perform.

Where there has been a gap in existing tools that help with sustainable design, Architype have developed their own. Ecco-lab is software that quickly assesses the cost and carbon impact from the earliest design stages, looking at low embodied energy and not just energy in use.

The tools are used in an iterative way to frame the brief requirements against the requirements of Passive House or other energy or carbon targets. Mapping this against the RIBA work stages indicates the layering of design development that occurs with Passive House projects. Using PHPP as an optimisation tool has consistently delivered some of the most energy-efficient buildings in the UK for both thermal and overall primary energy (including unregulated energy loads), as proven by post-occupancy monitoring.

Energy modelling is essential for an in-depth understanding of how much energy a building will use, but it's important to get the basics right too. There's a dynamic and sometimes difficult balance of orientation, form, building fabric, lifecycle impact, materials, services strategy, context, simple and elegant detailing, and climatic conditions such as daylight, ventilation and passive solar gain, which all need to be considered to deliver a truly successful building – and it shouldn't stop there. Architype monitor how their buildings perform through post-occupancy research, staying with a project to ensure building users know how to operate it for optimum performance, and applying the lessons learned to improve future designs.

Architype's accumulated knowledge across projects is shared through their online design collaboration tool and in person, both inside the practice and with the wider industry. Each project team includes experienced architects and staff trained in using PHPP modelling, who guide the schematic design. In design and technical reviews people outside of the core team advise on the implications of a design strategy. Initial designs are tested to establish that key moves – massing, orientation, site position and glazing – will achieve the Passive House Standard. Using a basic PHPP model, these early critical checks ensure the correct heating loads and primary energy targets are met. A Passive House workflow task list that correlates with the RIBA work stages is used as the project progresses. Constant optimisation takes place during the design process as more detailed data is entered into the model. Architype disseminate knowledge internally and externally, including sharing strategies that have been proven to work from post-occupancy monitoring. Robust energy reporting and feedback on massing and energy impacts from RIBA stages 1 and 2 up to construction information is shared with clients and the design team. The architectural and energy intentions of the scheme are

Figure 1:
Architype's technical associate Robert White sharing Passive House construction research with colleagues

continued through from inception to completion. There is joined-up thinking from the big idea through to the 1:2 detail of a window frame and its relationship to the reveal.

In some circumstances designers can repeat a known formula with small changes and revisions, such as delivering affordable Passive House schools for very tight budgets, as outlined in the Ysgol Bro Hyddgen school example (see below).

There are many reasons to be hopeful about the future. Architype are seeing a lot of client interest in using Passive House as the best route to achieve healthy buildings with good air quality, as well as the recent moves to account for embodied carbon in Passive House assessment. With more knowledge gained, confidence in what can be achieved has grown. The practice have already delivered the first Passive House Plus non-domestic building in the UK (an eco-centre in Bicester, see page 75), are building a zero-carbon school in Hackbridge, and are not far off achieving a Passive House Premium building that will produce more energy than it consumes. Investment in battery storage and other technologies that could make ordinary homes become energy generators themselves opens up even more possibilities.

Ysgol Bro Hyddgen
Passive House school in UK

Client: Powys Council/Dawnus Construction
Design team: WSP, Churchmans Landscaping
Text: Ann Marie Fallon, Associate and CEPH Designer

Figure 2:
Ysgol Bro Hyddgen main entrance approach from the east, showing the conceptual massing of Early Years through to the remaining massing

Ysgol Bro Hyddgen school is a proposed all-through school for the town of Machynlleth in Wales. The design is aiming to be the first Passive House all-through school in the UK, with provision for Early Years through to sixth-form. The site is optimised to allow this, having a southerly aspect. Building on this via the conceptual massing, the 'path' through the school from Early Years is reflected in the massing, with the younger pupils introduced to smaller-scale single-storey spaces, over time moving through to the main three-storey sixth-form spaces. A central theatre space, dining/social area and large sports hall comprise the central 'heart' space.

Figure 3:
Massing of ground floor plan on site, illustrating the southerly aspect and site shading to east and south

Sketches were developed to optimise the school design using established details, and inform early-stage PHPP modelling and BIM Revit modelling (thermal bridges, project boundaries, etc.).

Figure 4:
Interim design review of sketch details, hand drawn for discussion to inform design intent, and also the BIM strategy in detail delivery

The additional users and emphasis on community use of the main hall theatre and gathering spaces give the scheme a good year-round use profile. As a result, this expanded typology will probably have a variety of users outside the normalised school hours, and primary energy loads will vary.

This study demonstrates practice development in relation to alignment of PHPP modelling and utilisation of 3D design tools within projects. It was proposed, via internal consultation with experienced Passive House designers in the office and our BIM champions, that there was a more efficient way to inform the data for the PHPP. This could be delivered by utilising the data in the BIM. Ysgol Bro Hyddgen was the first project on which this method was trialled, and it is under review on a case-by-case basis, dependent on the scale and scheme complexity. The design team worked closely with Passive House certifiers to ensure that the methodology worked and would remain acceptable to the Passivhaus Institute, which would ultimately certify the building.

Taking data from the BIM, templates were developed to display accurate values of each area in the school, which in turn created a simpler and more legible visual, accessible by people not trained in Passive House design. This also enabled the tasks around Passive House evidencing and design development assessment to be more accessible to non-Passive House designers in the practice, which further improved workflow and resource capacity. This informal upskilling by implementing BIM principles and aligning PHPP with these in a simplified manner optimises the workflow efficiencies given the increased scale and complexity of Passive House projects across the practice.

Figure 5:
CNC model developed in-house by Architype for public consultations on the school

PHPP Roof Areas Schedule			
Type Mark	Area	Mark	Description
PHRTF-P	495.5 m²	P	Roof to Primary (Early part - 1 storey block)
PHRTF-C	1920.0 m²	C	Roof to Central Block (2 storey block)
PHRFT-SP	669.2 m²	SP	Roof to Sports Hall
PHRTF-S	954.0 m²	S	Roof to Secondary Block (3 storey block)
PHRFT-SP	6.6 m²	LO	Roof to Lift overun
PHRFT-SP	53.0 m²	S_CL2	Roof to Secondary block Clerestory2
PHRFT-SP	62.5 m²	S_CL1	Roof to Secondary block Clerestory1
PHRFT-SP	56.1 m²	C_CL1	Roof to Central Block Hub (2 storey block)
PHRFT-SP	82.2 m²	P_CL2	Roof to Primary Hall Clerestory
PHRFT-SP	49.1 m²	P_CL1	Roof to Primary Hub Clerestory

PHPP Elevation Areas Schedule				
X	Area	Mark	Comments	Rotation From True North
PHWT-E	6.9 m²	C_CL_E	Central Block – Hub Clerestory	57.91°
PHWT-N	27.3 m²	C_CL_N	Central Block – Hub Clerestory	327.91°
PHWT-S	37.0 m²	C_CL_S	Central Block – Hub Clerestory	147.91°
PHWT-W	6.9 m²	C_CL_W	Central Block – Hub Clerestory	237.91°
PHWT-E	197.9 m²	C_E1	Central Block – East Wall (Main Entrance)	57.91°
PHWT-E	193.9 m²	C_E2	Central Block – East Wall (Sports Hall)	57.91°
PHWT-N	308.8 m²	C_N	Central Block – North Wall	327.91°
PHWT-S	88.4 m²	C_S	Central Block – South Wall	147.91°
PHWT-S	2.1 m²	C_S-1	Central Block – South Wall (Bellow Slab)	237.91°
PHWT-N	493.9 m²	C_S_N	Central Block – North Wall	327.91°
PHWT-W	303.7 m²	C_W	Central Block – West Wall	237.91°
PHWT-E	3.1 m²	LO_E	Secondary Block – Liftoverun	57.91°
PHWT-N	3.3 m²	LO_N	Secondary Block – Liftoverun	327.91°
PHWT-S	3.3 m²	LO_S	Secondary Block – Liftoverun	147.91°
PHWT-W	3.1 m²	LO_W	Secondary Block – Liftoverun	237.91°
PHWT-E	8.4 m²	P_CL1_E	Primary Block – Hub Clerestory	57.91°
PHWT-N	16.1 m²	P_CL1_N	Primary Block – Hub Clerestory	327.91°
PHWT-S	24.6 m²	P_CL1_S	Primary Block – Hub Clerestory	147.91°
PHWT-W	8.4 m²	P_CL1_W	Primary Block – Hub Clerestory	237.91°
PHWT-E	8.3 m²	P_CL2_E	Primary Block – Hall Clerestory	57.91°
PHWT-N	29.2 m²	P_CL2_N	Primary Block – Hall Clerestory	327.91°
PHWT-S	43.5 m²	P_CL2_S	Primary Block – Hall Clerestory	147.91°
PHWT-W	8.3 m²	P_CL2_W	Primary Block – Hall Clerestory	237.91°
PHWT-E	77.7 m²	P_E1	Primary Block – East Wall (Early Years 1-storey)	57.91°
PHWT-E	94.2 m²	P_E2	Primary Block – East Wall (2-storey)	57.91°
PHWT-N	105.3 m²	P_N1	Primary Block – North Wall (Early Years 1-storey)	327.91°
PHWT-N	162.0 m²	P_N2	Primary Block – North Wall (Primary Hall 2-storey)	327.91°
PHWT-S	387.7 m²	P_S1	Primary Block – South Wall	147.91°
PHWT-S	140.5 m²	P_S2	Primary Block – South Wall (Dining Atrium)	147.91°
PHWT-W	61.2 m²	P_W1	Primary Block – West Wall (Stairwell)	237.91°
PHWT-W	46.5 m²	P_W2	Primary Block – West Wall (Dining Atrium)	237.91°
PHWT-E	8.9 m²	S_CL1_E	Secondary Block – Hub1 Clerestory	57.91°
PHWT-N	24.0 m²	S_CL1_N	Secondary Block – Hub1 Clerestory	327.91°
PHWT-S	34.9 m²	S_CL1_S	Secondary Block – Hub1 Clerestory	147.91°
PHWT-W	8.9 m²	S_CL1_W	Secondary Block – Hub1 Clerestory	237.91°
PHWT-E	8.9 m²	S_CL2_E	Secondary Block – Hub2 Clerestory	57.91°
PHWT-N	20.4 m²	S_CL2_N	Secondary Block – Hub2 Clerestory	327.91°
PHWT-S	29.6 m²	S_CL2_S	Secondary Block – Hub2 Clerestory	147.91°
PHWT-W	8.9 m²	S_CL2_W	Secondary Block – Hub2 Clerestory	237.91°
PHWT-E	66.4 m²	S_E	Secondary Block – East Wall	57.91°
PHWT-S	474.9 m²	S_S1	Secondary Block – South Wall (Main facade)	147.91°
PHWT-S	21.0 m²	S_S2	Secondary Block – South Wall (Small stepback)	147.91°
PHWT-W	84.5 m²	S_W1	Secondary Block – West Wall	237.91°
PHWT-W	133.7 m²	S_W2	Secondary Block – West Wall (Stairwell)	237.91°
	3826.4 m²			

PHPP Floor Areas Schedule			
WT	Area	Mark	Description
PHEFT-C	1218.4 m²	C	Floor to Central Block
PHEFT-C	646.2 m²	SH	Floor to Sports Hall
PHEFT-P	1324.5 m²	P	Floor to Primary Block
PHEFT-S	847.0 m²	S	Floor to Secondary Block
Grand total: 4	4036.1 m²		

Floor Slab Perimeter: 387.47m

2 PHPP - Roofs

3 PHPP - Floors

1 1 external 2

Figure 6:
Highlighting key elevation and areas for the Passive House designer and the certifier for optimised evidencing (heat-loss areas)

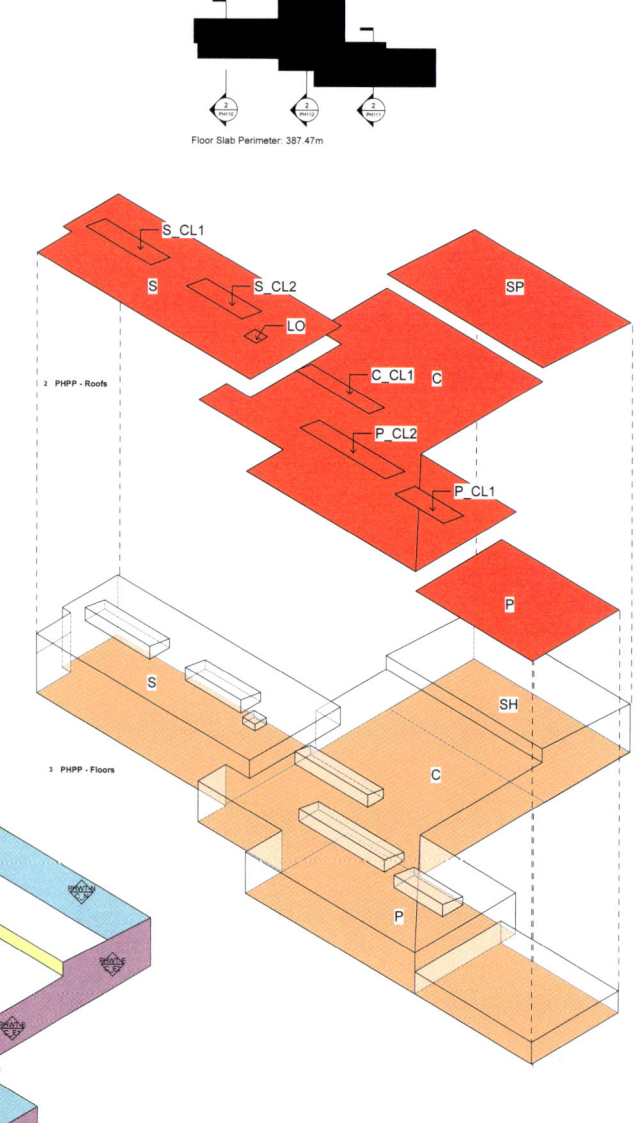

TFA Schedule GF

Number	Name	Department	Area	TFA Schedule	TFA Percentage	TFA	Subtractions	final tfa
00 GF								
00.01	Lobby	Auxillary & WCs	8.0 m²	Useful Areas	100	8.0 m²		
00.02	Reception	Auxillary & WCs	13.7 m²	Useful Areas	100	13.7 m²		
00.03	Circulation	Circulation	17.7 m²	Access Areas	60	10.6 m²		
00.04	ACC WC	Auxillary & WCs	3.6 m²	Useful Areas	100	3.6 m²		
00.05	Circulation	Circulation	30.6 m²	Access Areas	60	18.4 m²		
00.06A	General Office	Staff & Admin	22.2 m²	Useful Areas	100	22.2 m²		
00.06B	Sick Bay	Staff & Admin	7.9 m²	Useful Areas	100	7.9 m²		
00.07	Learning Resources	Learning Resource	54.8 m²	Useful Areas	100	54.8 m²	0.0 m²	54.8 m²
00.08	Stair 1	Circulation	10.6 m²	Vertical Circulation	0	0.0 m²	2.0 m²	-2.0 m²
00.09	Caretaker Store	Stores	5.4 m²	Useful Areas	100	5.4 m²		
00.10	Caretaker's Office	Staff & Admin	10.3 m²	Useful Areas	100	10.3 m²		
00.11	Small Group & Interview	Learning Resource	10.3 m²	Useful Areas	100	10.3 m²		
00.12	SEN / WChair St	Stores	12.2 m²	Useful Areas	100	12.2 m²		
00.13	Hygiene Room 2	Auxillary & WCs	12.2 m²	Useful Areas	100	12.2 m²		
00.14	Circulation	Circulation	81.6 m²	Access Areas	60	49.0 m²		
00.14A	Community Lockers	Stores	3.5 m²	Useful Areas	100	3.5 m²		
00.15	PE St	Stores	30.4 m²	Useful Areas	100	30.4 m²		
00.16	PE St	Stores	22.5 m²	Useful Areas	100	22.5 m²		
00.17	4 Court Sports Hall	Hall	600.9 m²	Useful Areas	100	600.9 m²		
00.18	PE Office	Staff & Admin	9.8 m²	Useful Areas	100	9.8 m²		
00.19	ACC/Staff Change	Auxillary & WCs	8.9 m²	Useful Areas	100	8.9 m²		
00.20A	Pupil Change & Shower	Auxillary & WCs	29.7 m²	Useful Areas	100	29.7 m²		
00.20B	WC	Stores	2.5 m²	Useful Areas	100	2.5 m²		
00.21A	Pupil Change & Shower	Auxillary & WCs	29.9 m²	Useful Areas	100	29.9 m²		
00.21B	WC	Stores	2.5 m²	Useful Areas	100	2.5 m²		
00.22	Maintenance St	Stores	6.8 m²	Useful Areas	100	6.8 m²		
00.23	Staging/ App. St	Stores	8.6 m²	Useful Areas	100	8.6 m²		
00.24	Cleaner	Auxillary & WCs	2.8 m²	Useful Areas	100	2.8 m²		
00.24A	Circulation	Circulation	137.8 m²	Access Areas	60	82.7 m²		
00.24B	Dining & Social Space	Hall	158.0 m²	Useful Areas	100	158.0 m²		
00.24C	Stair 2	Circulation	36.1 m²	Vertical Circulation	0	0.0 m²		
00.24D	Dining & Social 6th Form	Hall	36.4 m²	Useful Areas	100	36.4 m²		
00.25A	Hall Lobby	Circulation	4.8 m²	Useful Areas	100	4.8 m²		
00.25B	School Hall	Hall	190.6 m²	Useful Areas	100	190.6 m²		
00.26A	Catering Kitchen	Auxillary & WCs	77.1 m²	Useful Areas	100	77.1 m²		

TFA Schedule GF

Number	Name	Department	Area	TFA Schedule	TFA Percentage	TFA	Subtractions	final tfa
00.26B	Kit St.	Auxillary & WCs	5.2 m²	Useful Areas	100	5.2 m²		
00.26C	Kit Office	Auxillary & WCs	3.2 m²	Useful Areas	100	3.2 m²		
00.26D	Kit Changing	Auxillary & WCs	2.5 m²	Useful Areas	100	2.5 m²		
00.26E	Kit WC	Auxillary & WCs	2.1 m²	Useful Areas	100	2.1 m²		
00.26F	Kit COSHH	Stores	2.6 m²	Useful Areas	100	2.6 m²		
00.27	Lift 1	Auxillary & WCs	2.9 m²	Vertical Circulation	0	0.0 m²		
00.28A	Pupil WCs 1	Auxillary & WCs	11.7 m²	Useful Areas	100	11.7 m²		
00.28B	Pupil WCs 2	Auxillary & WCs	11.1 m²	Useful Areas	100	11.1 m²		
00.29A	Circulation	Circulation	78.0 m²	Access Areas	60	46.8 m²		
00.29B	Circulation	Circulation	77.3 m²	Access Areas	60	46.4 m²		
00.30	Stair 3	Circulation	15.9 m²	Vertical Circulation	0	0.0 m²		
00.31	Technology Metalwork	Teaching	92.4 m²	Useful Areas	100	92.4 m²		
00.32	Resistant Mat Prep	Stores	37.8 m²	Useful Areas	100	37.8 m²		
00.34	Cleaner	Stores	3.8 m²	Useful Areas	100	3.8 m²		
00.35	ACC WC/Staff	Auxillary & WCs	4.8 m²	Useful Areas	100	4.8 m²		
00.36	Technology Woodwork	Teaching	79.5 m²	Useful Areas	100	79.5 m²		
00.37	ALN & Kitchenette	Learning Resource	56.6 m²	Useful Areas	100	56.6 m²		
00.38	Stair 4	Circulation	15.9 m²	Vertical Circulation	0	0.0 m²		
00.39	Cookery Room	Teaching	90.5 m²	Useful Areas	100	90.5 m²		
00.40	2D Art	Teaching	93.6 m²	Useful Areas	100	93.6 m²		
00.41	Art & Design Resources	Learning Resource	43.4 m²	Useful Areas	100	43.4 m²		
00.42	Textiles	Teaching	44.6 m²	Useful Areas	100	44.6 m²		
00.43	Elec Cupbd	Auxillary & WCs	5.0 m²	Useful Areas	100	5.0 m²		
00.46A	Circulation	Circulation	34.8 m²	Access Areas	60	20.9 m²		
00.46B	Circulation	Circulation	73.9 m²	Access Areas	60	44.4 m²		
00.46C	Library Resource	Learning Resource	33.2 m²	Useful Areas	100	33.2 m²		
00.46D	Food Science & DT	Teaching	38.1 m²	Useful Areas	100	38.1 m²		
00.47	Stair 5	Circulation	14.5 m²	Vertical Circulation	0	0.0 m²		
00.48	Class 2	Teaching	60.9 m²	Useful Areas	100	60.9 m²		
00.49	Class 1	Teaching	61.3 m²	Useful Areas	100	61.3 m²		
00.50	Reception 2	Teaching	68.2 m²	Useful Areas	100	68.2 m²		
00.51	WCs	Auxillary & WCs	6.5 m²	Useful Areas	100	6.5 m²		
00.52	Reception 1	Teaching	68.1 m²	Useful Areas	100	68.1 m²		
00.53	WCs	Auxillary & WCs	6.4 m²	Useful Areas	100	6.4 m²		
00.54	Small Grp Rm	Learning Resource	9.4 m²	Useful Areas	100	9.4 m²		
00.55	ALN & Kitchenette	Learning Resource	32.1 m²	Useful Areas	100	32.1 m²		
00.56	Stair 6	Circulation	19.1 m²	Vertical Circulation	0	0.0 m²		
00.57	Lift 2	Circulation	3.9 m²	Vertical Circulation	0	0.0 m²		
00.58	Hygiene Room 1	Auxillary & WCs	12.2 m²	Useful Areas	100	12.2 m²		
00.59A	Reception	Staff & Admin	12.3 m²	Useful Areas	100	12.3 m²		

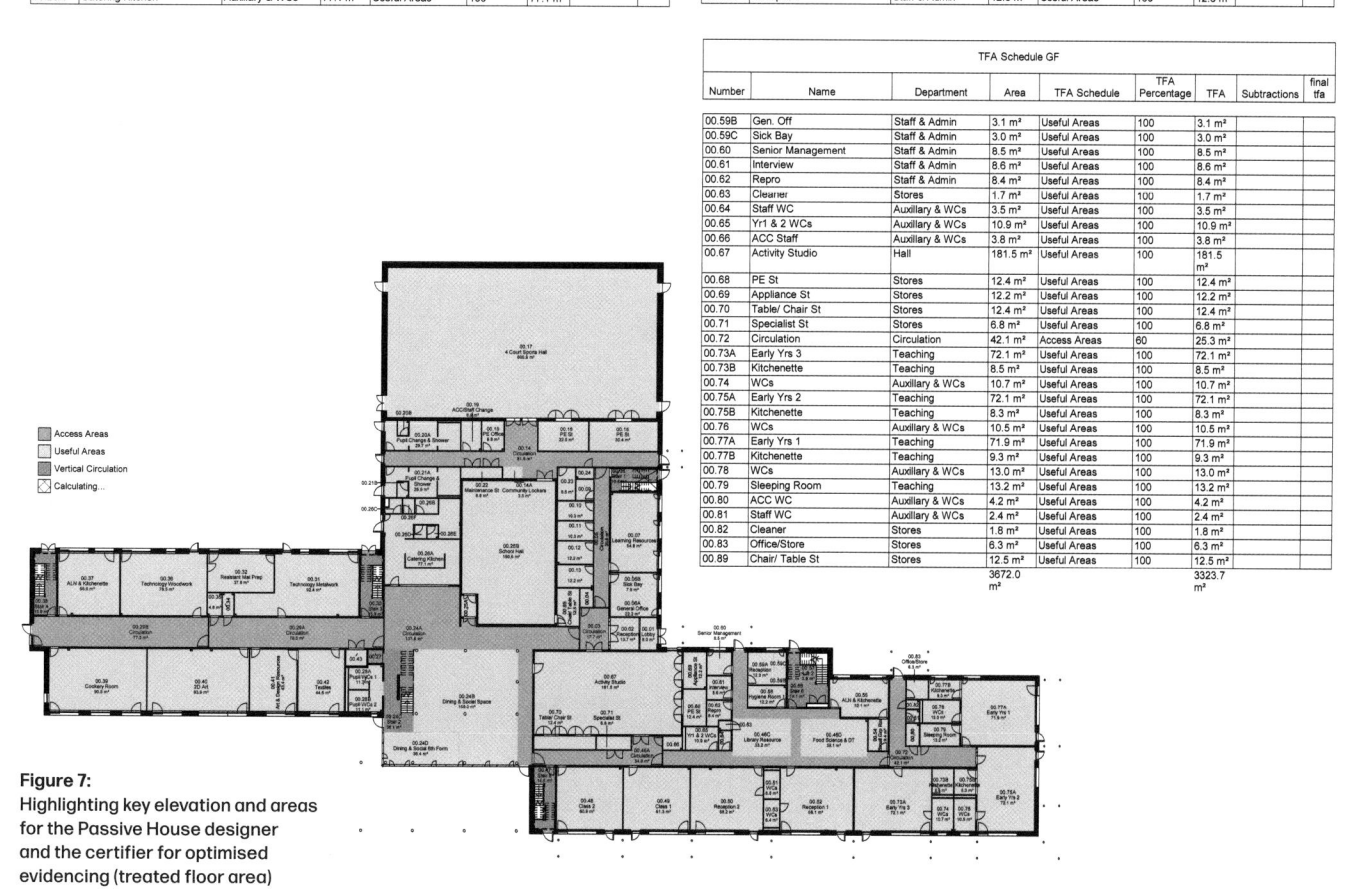

TFA Schedule GF

Number	Name	Department	Area	TFA Schedule	TFA Percentage	TFA	Subtractions	final tfa
00.59B	Gen. Off	Staff & Admin	3.1 m²	Useful Areas	100	3.1 m²		
00.59C	Sick Bay	Staff & Admin	3.0 m²	Useful Areas	100	3.0 m²		
00.60	Senior Management	Staff & Admin	8.5 m²	Useful Areas	100	8.5 m²		
00.61	Interview	Staff & Admin	8.6 m²	Useful Areas	100	8.6 m²		
00.62	Repro	Staff & Admin	8.4 m²	Useful Areas	100	8.4 m²		
00.63	Cleaner	Stores	1.7 m²	Useful Areas	100	1.7 m²		
00.64	Staff WC	Auxillary & WCs	3.5 m²	Useful Areas	100	3.5 m²		
00.65	Yr1 & 2 WCs	Auxillary & WCs	10.9 m²	Useful Areas	100	10.9 m²		
00.66	ACC Staff	Auxillary & WCs	3.8 m²	Useful Areas	100	3.8 m²		
00.67	Activity Studio	Hall	181.5 m²	Useful Areas	100	181.5 m²		
00.68	PE St	Stores	12.4 m²	Useful Areas	100	12.4 m²		
00.69	Appliance St	Stores	12.2 m²	Useful Areas	100	12.2 m²		
00.70	Table/ Chair St	Stores	12.4 m²	Useful Areas	100	12.4 m²		
00.71	Specialist St	Stores	6.8 m²	Useful Areas	100	6.8 m²		
00.72	Circulation	Circulation	42.1 m²	Access Areas	60	25.3 m²		
00.73A	Early Yrs 3	Teaching	72.1 m²	Useful Areas	100	72.1 m²		
00.73B	Kitchenette	Teaching	8.5 m²	Useful Areas	100	8.5 m²		
00.74	WCs	Auxillary & WCs	10.7 m²	Useful Areas	100	10.7 m²		
00.75A	Early Yrs 2	Teaching	72.1 m²	Useful Areas	100	72.1 m²		
00.75B	Kitchenette	Teaching	8.3 m²	Useful Areas	100	8.3 m²		
00.76	WCs	Auxillary & WCs	10.5 m²	Useful Areas	100	10.5 m²		
00.77A	Early Yrs 1	Teaching	71.9 m²	Useful Areas	100	71.9 m²		
00.77B	Kitchenette	Teaching	9.3 m²	Useful Areas	100	9.3 m²		
00.78	WCs	Auxillary & WCs	13.0 m²	Useful Areas	100	13.0 m²		
00.79	Sleeping Room	Teaching	13.2 m²	Useful Areas	100	13.2 m²		
00.80	ACC WC	Auxillary & WCs	4.2 m²	Useful Areas	100	4.2 m²		
00.81	Staff WC	Auxillary & WCs	2.4 m²	Useful Areas	100	2.4 m²		
00.82	Cleaner	Stores	1.8 m²	Useful Areas	100	1.8 m²		
00.83	Office/Store	Stores	6.3 m²	Useful Areas	100	6.3 m²		
00.89	Chair/ Table St	Stores	12.5 m²	Useful Areas	100	12.5 m²		
			3672.0 m²			3323.7 m²		

Figure 7:
Highlighting key elevation and areas for the Passive House designer and the certifier for optimised evidencing (treated floor area)

Archetype - a question of culture

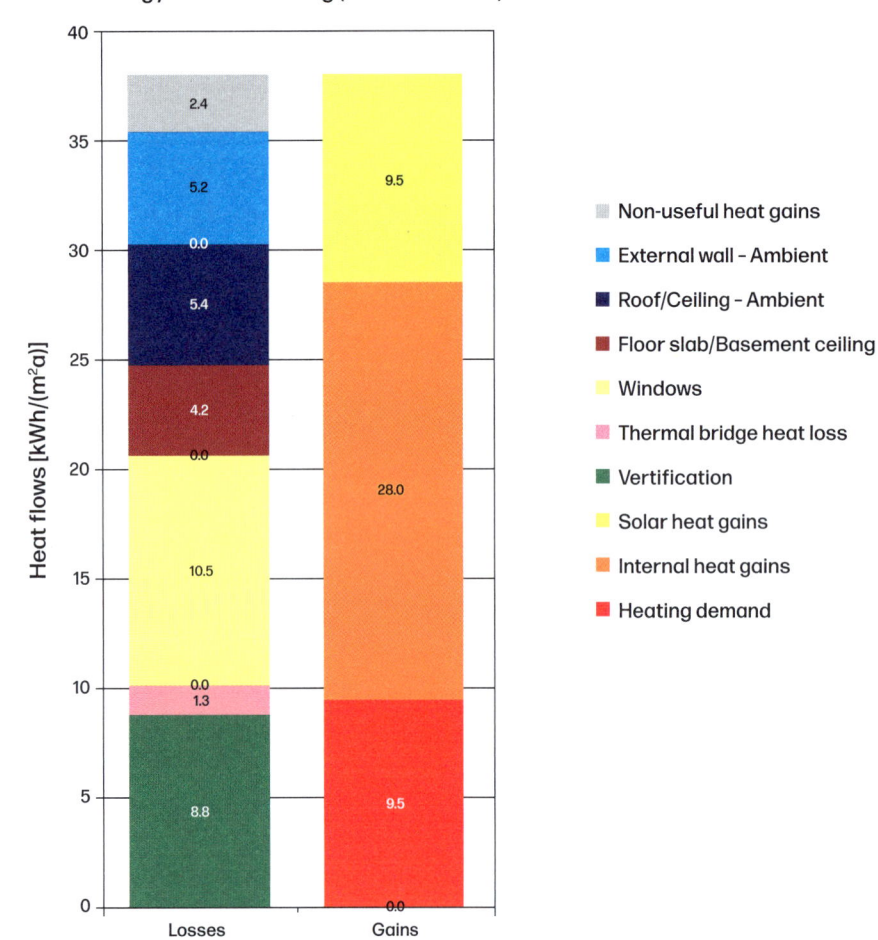

Figure 8:
Enabling easier assessment of energy balances in complex projects by optimising inputs into PHPP

Energy balance heating (annual method)

Losses:
- Non-useful heat gains: 2.4
- External wall – Ambient: 5.2
- Roof/Ceiling – Ambient: 5.4
- Floor slab/Basement ceiling: 4.2
- Windows: 10.5
- Thermal bridge heat loss: 1.3
- Ventilation: 8.8

Gains:
- Solar heat gains: 9.5
- Internal heat gains: 28.0
- Heating demand: 9.5

In this trial of coordinating BIM and PHPP elements, most of the information translated well. However, in some cases more customised algorithms were needed for the bespoke areas of the project, such as different typologies when applying this to future projects. There are a growing number of Revit-based PHPP BIM packages becoming available, but it was important to the practice that a simple interface was developed. This was to allow the basic heat-loss and glazing fundamentals to be easily accessible to less-experienced Passive House designers, with more experienced PHPP practitioners then able to focus on MEP and other complexities to be resolved in the design further on.

Energy Modelling in Architecture

Bicester Eco Business Centre

Client: Cherwell District Council
Design team: E3 MEP Engineers
Text: Rebecca Robinson, Architectural Assistant and CEPH Designer; Ann-Marie Fallon, Associate and CEPH Designer

Bicester Eco Business Centre, just outside of Oxford and completed in December 2018, is the first commercial building to be certified to the Passive House Plus Standard in the UK. The building sits within the zero-carbon Bicester Eco Town, and creates employment for 125 people in a 'co-working collaborative business incubator', aimed at local and start-up businesses.

The brief was for a zero-carbon in use, BREEAM Excellent, flagship building. Architype proposed to the client that, given the other standards being targeted, Passive House Plus could also be possible. Passive House Plus is an enhanced version of the classic Passive House Standard. It includes the standard low-energy heating demand criteria of Passive House but has more stringent criteria for primary energy, which is balanced by renewable energy generation, making the building essentially net-zero energy in terms of operational energy use.

Figure 9:
Form evolution from masterplan intent to final developed concept

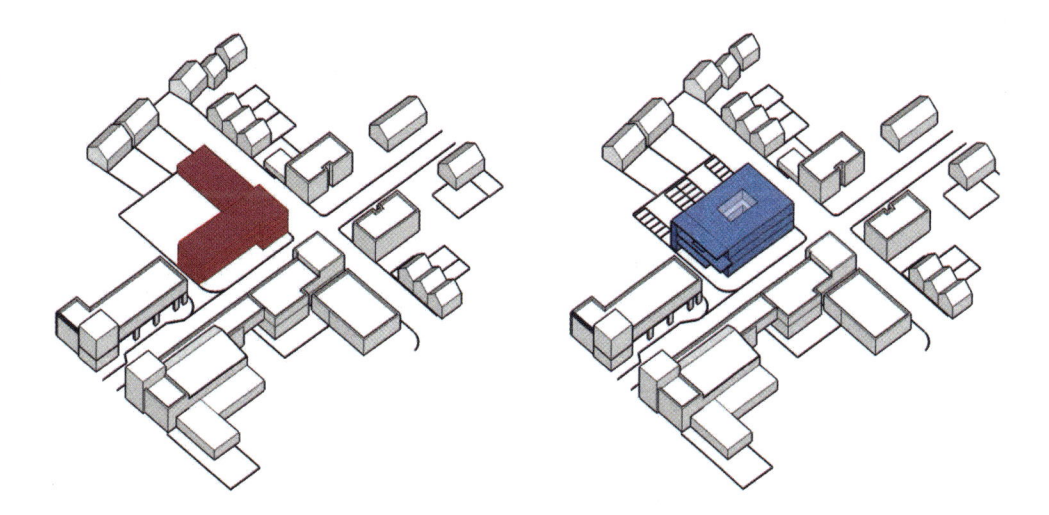

The early concept design was a good example of an in-house Architype-developed approach to massing, form and orientation. While the outline planning scheme for the building showed a thin, L-shaped building, the internal team evolved a more compact form. This increased the building's potential to reach Passive House cost at an optimised specification. The compact form and deeper plan also retained good daylighting and enabled efficient circulation via a central top-lit atrium. This central space also became the communal 'hub' of the building.

Figure 10:
Concrete and timber infill coming together from concept to site

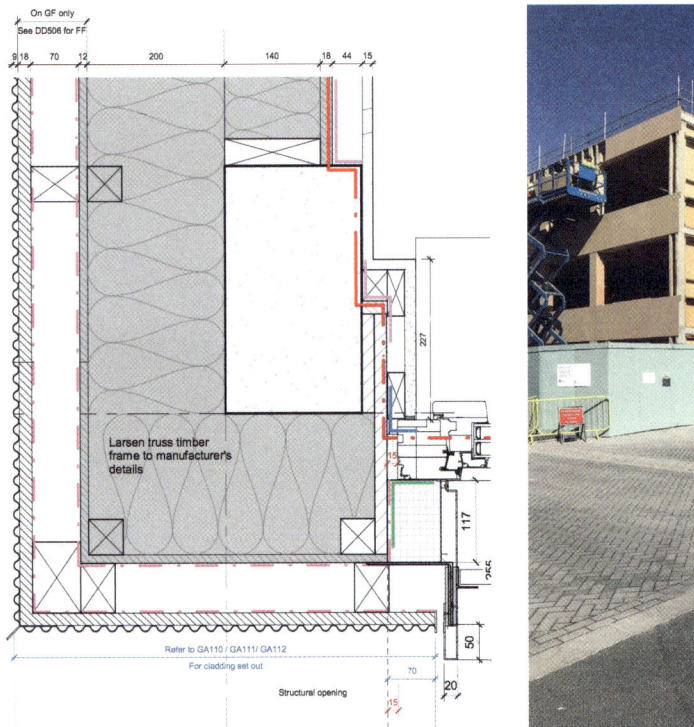

The structure was concrete frame. In order to compensate in a small way for the embodied energy of using a concrete frame, the option of timber frame infill (versus steel) was chosen. An external skin of timber fins encases the building to control solar gain, as well as being a central design device to soften the building and give some semi-private terrace access to the users of the building. The materiality of the building was kept very simple: metal sheet main walls clad in a lighter timber skin.

Figure 11:
Concept of atrium-based massing clad in a 'second skin' of timber

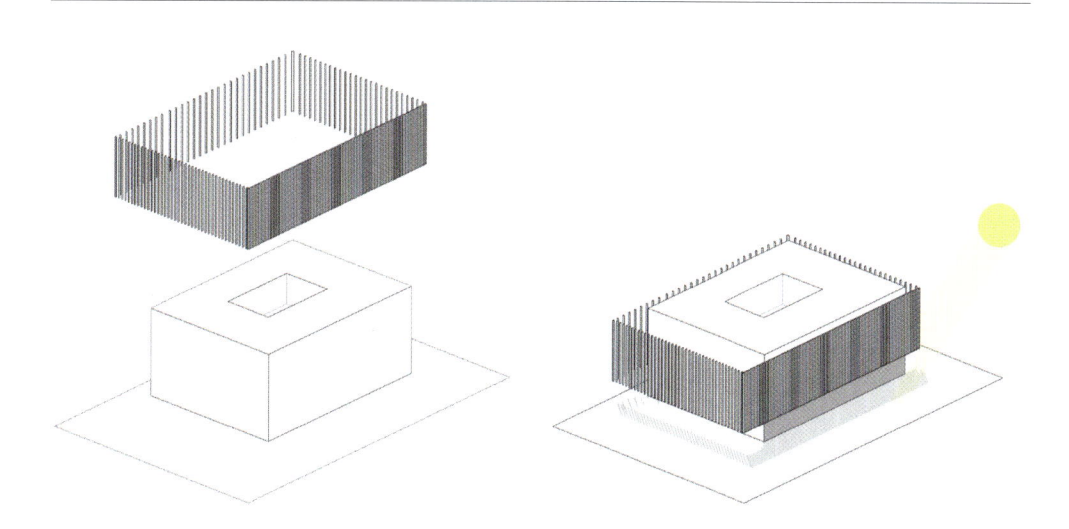

Architype commissioned in-depth shading analysis in order to understand the impact of both solar gain in winter and overheating in summer. Bespoke timber cladding was developed to mitigate solar gain from the south, and horizontal brise-soleils were also used to shade the high summer sun, given that vertical shading is less successful to the south in the UK. To the east and west, the vertical fins shade against the oblique angle of the sun, with an additional external walkway on the east side to mark the entrance to the building.

Figure 12:
Early design parti sketches integrating the shading/orientation analysis with the development of architectural language/intent

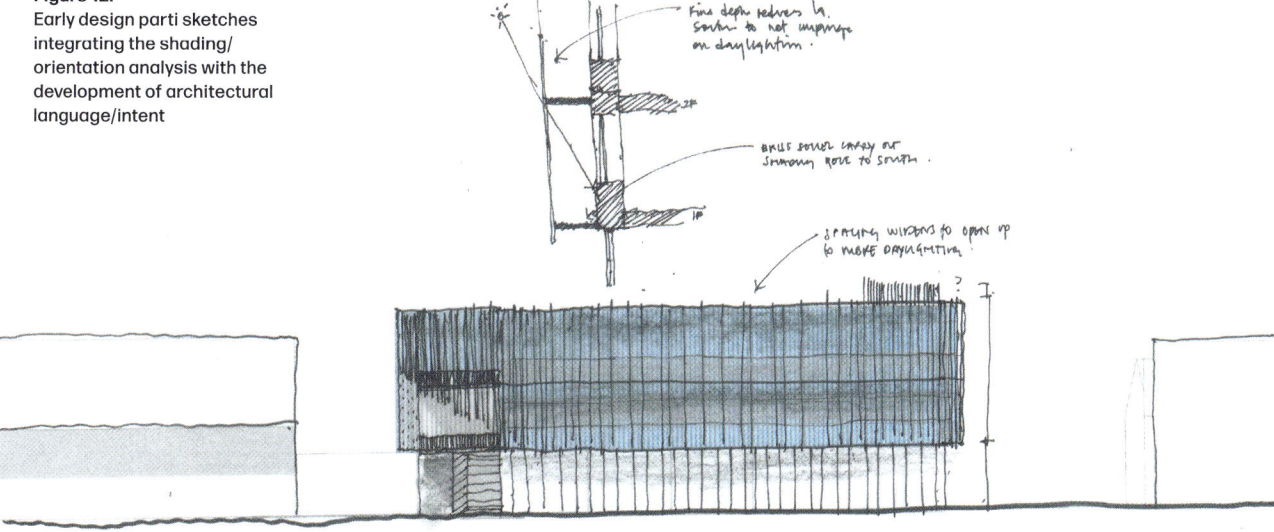

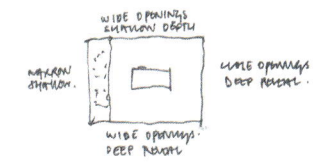

Figure 13:
External shading devices

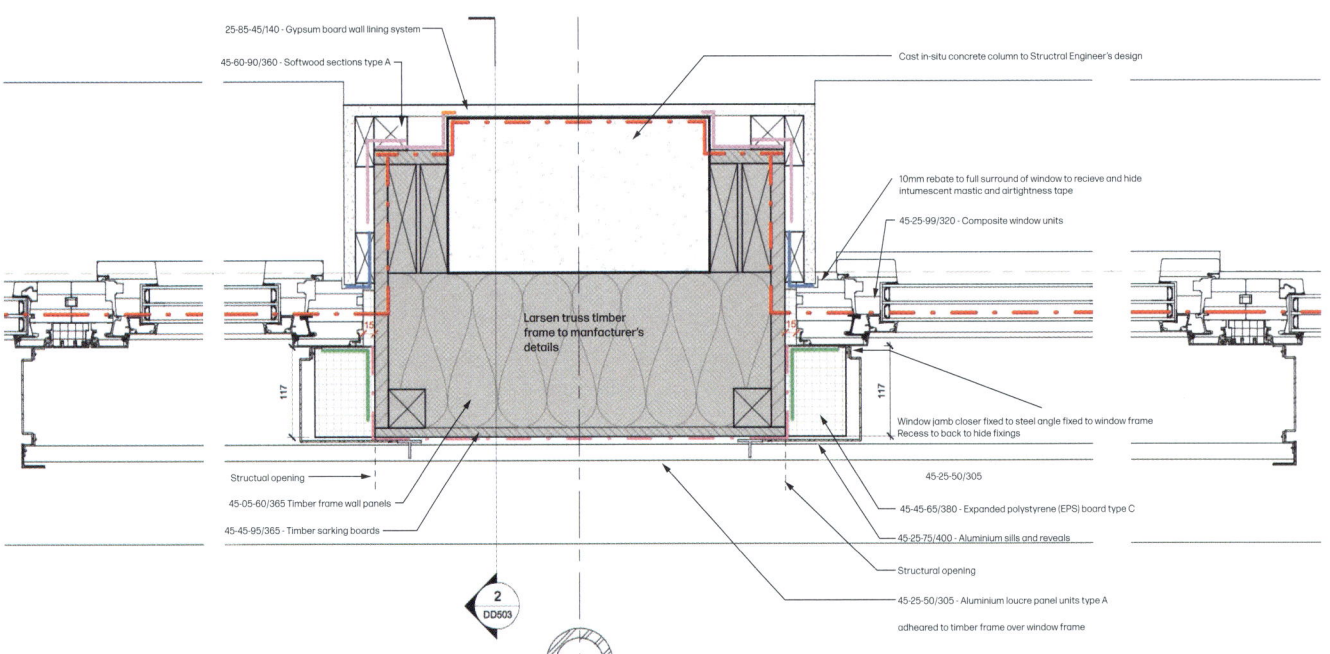

External roller shading on the ground floor was also installed to balance security, access and legibility of the ground floor in architectural terms.

The process of modelling the building in PHPP means that designers can drill down into all elements of the building and focus on optimising them in terms of energy and design, even down to the number of hot water taps, or the number of IT appliances including Ethernet points, as well as extensive analysis of the building fabric.

76   Energy Modelling in Architecture

PHPP also demands true interrogation of the details, component by component, to record and optimise their thermal bridging impact. Figure 14 is a detail developed to fix the timber steel frame secondary structure to the main concrete frame. The connections needed to be thermally broken, as each connection is accounted for in the heat-loss section of the PHPP modelling. This is assessed in 2D and 3D and informs design intent; it is referred to constantly right the way through to subcontractor sign-off and when viewing the installation on site as evidence collation for Passive House certification.

Figure 14:
Working through complex fixing strategies and ensuring their thermal impact is accounted for in PHPP

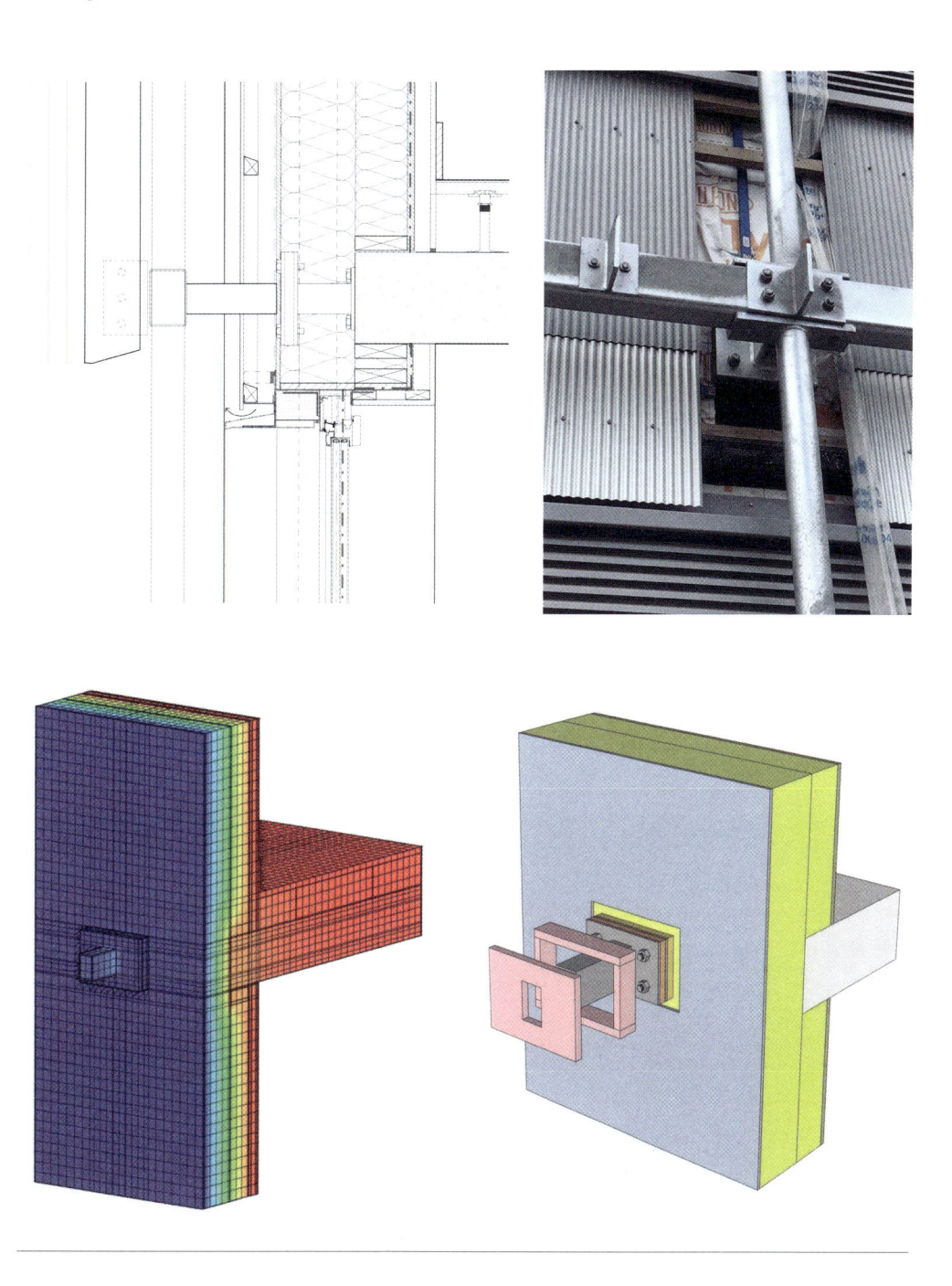

78   Energy Modelling in Architecture

Figure 15:
External and internal views

When assessing energy use in the building, primary energy had to be optimised to achieve the stringent criteria for Passive House Plus ('primary energy', in relation to Passive House, is defined as the total operational energy used by the building). This consists of regulated energy such as space heating and hot water, and unregulated energy such as plugged-in devices and appliances. A factor is added to account for the losses for particular energy sources, i.e. gas versus mains electricity. The future de-carbonisation of the grid towards 2050 will start establishing similar energy factors for gas and renewable electricity, for example. Passive House Plus offers a way of considering this now.

Detailed analysis of all equipment in the building was undertaken in relation to utilisation factors, hours of use and power/energy ratings. A process of optimisation followed – either more efficient appliances, more efficient control, or omission/ simplification where possible. This process forces the designer to interrogate the need for each item in the building rather than simply accepting service/appliance information. For example, hand dryers were replaced with reusable towels and paper towels, and the need for several printers was removed by having one large printer in a centralised repro room on the ground floor. This also encouraged more interaction between occupants, an important aspect of the brief. The remaining appliances then needed to have the highest level of energy efficiency. It was clear that to get to this high level of energy performance, the client and users played an important part in the design decision-making. In this case the users are business start-ups with a focus on sustainability, so having one large printer, for instance, was not a hard sell due to their existing ethos on sustainability as a whole.

Large firms
4

Part 4 consists of discussions and case studies from three large firms: AHMM, Feilden Clegg Bradley (FCB) Studios and HOK. In all discussions, importance is placed on developing bespoke in-house approaches that enable a distinct design ethos across and between teams and multiple international and national office locations. **AHMM** reflect on their development of an in-house building performance team and sustainability programme, drawing out bespoke measures adopted across projects and teams. **FCBStudios** discuss the importance of maintaining a shared ethos and vision across multiple offices and teams through active supervision, as well as an overarching atmosphere of sustainability. **HOK** also observe the challenges of multiple offices and locations, reflecting on the need for sustainability leadership and motivation.

All three firms highlight the need to develop and maintain a shared ethos across multiple locations and project teams.

Chapter 8: AHMM – from programme to practice

AHMM's sustainability programme has developed organically, initially coupling in-house enthusiasm with an external consultant to advise design teams on an ad-hoc basis. As projects grew in scale and complexity and the regulatory framework surrounding sustainability required more input and detailed analysis, expertise was brought in-house with a sustainability specialist in 2014. This internal resource has since grown to a four-strong building performance team (BPT) led by a head of sustainability (HoS). The team is an integral part of AHMM's architectural design process, supporting designers with technical expertise and performance modelling. Formal project reviews are led by the HoS, where opportunities to test an architectural proposal through building performance modelling are identified. Modelling helps test ideas and prove concepts, develops integrated designs and ultimately helps deliver architecture that optimises both building performance and clients' investment.

Practice and projects

Allford Hall Monaghan Morris (AHMM) comprises multiple architectural teams organised in director-run studios led by four directors based in London. Each studio is comprised of associate directors, associates, project architects, senior architects, architects and architectural assistants. In mid-2019, the practice has just under 500 staff – around 400 architectural staff and 100 support staff. The practice is spread across three offices, with 440 people in London, 50 in Bristol and ten in Oklahoma City.

The BPT operates as part of a larger technical group as an internal consultancy, influencing massing, envelopes, fabric, details and specifications. This feeds into the information produced to procure buildings through drawings, specifications and written reports.

The concept of sustainability in architecture has undergone several transformations over recent years, from a niche design approach to a fully understood cross-industry imperative to an assumed completed task that can be taken for granted as assimilated into design and construction working practices.

Meanwhile BREEAM, the defunct Code for Sustainable Homes, other certification schemes and even Approved Document L have made some specialist thinking and technologies mainstream and have created some value in the idea of a 'green' building. However, these certification methods straddle a line between commercially acceptable tinkering and boundary-pushing challenge to the status quo. They do not necessarily represent the cutting edge of 'sustainable' design and in some cases have generated a false sense of achievement. As an industry, we think we are doing better than we are. The many studies comparing a Building Regulations energy calculation versus actual performance in use illustrate the weaknesses in regulations, our industry's inability to deliver to energy standards, and the potential pitfalls of any complacency.

Occupant wellbeing is the current commercially potent design driver, often leaving sustainability and energy performance to be considered as secondary aspects. While the wellbeing agenda represents a powerful lever that can help deliver a lower-impact built environment, the focus is not on CO_2. Industry must not lose sight of the scientific imperative to deal with climate change brought so effectively back into focus by the Extinction Rebellion protests in London and around the world.

This shifting market forms the context for AHMM's drive to deliver low-environmental-impact buildings. The continually changing targets that are mapped and then occasionally scrapped (through the Deregulation Bill, for example) and then reinstated by other means (the GLA has found a way of reinstating a zero-carbon building standard; the UKGBC has published a methodology for net-zero-carbon development) creates a changing legislative landscape. Constantly improving targets are designed to give industry time to adapt and develop the skills and knowledge required to deliver

Figure 1:
Practice oranogram

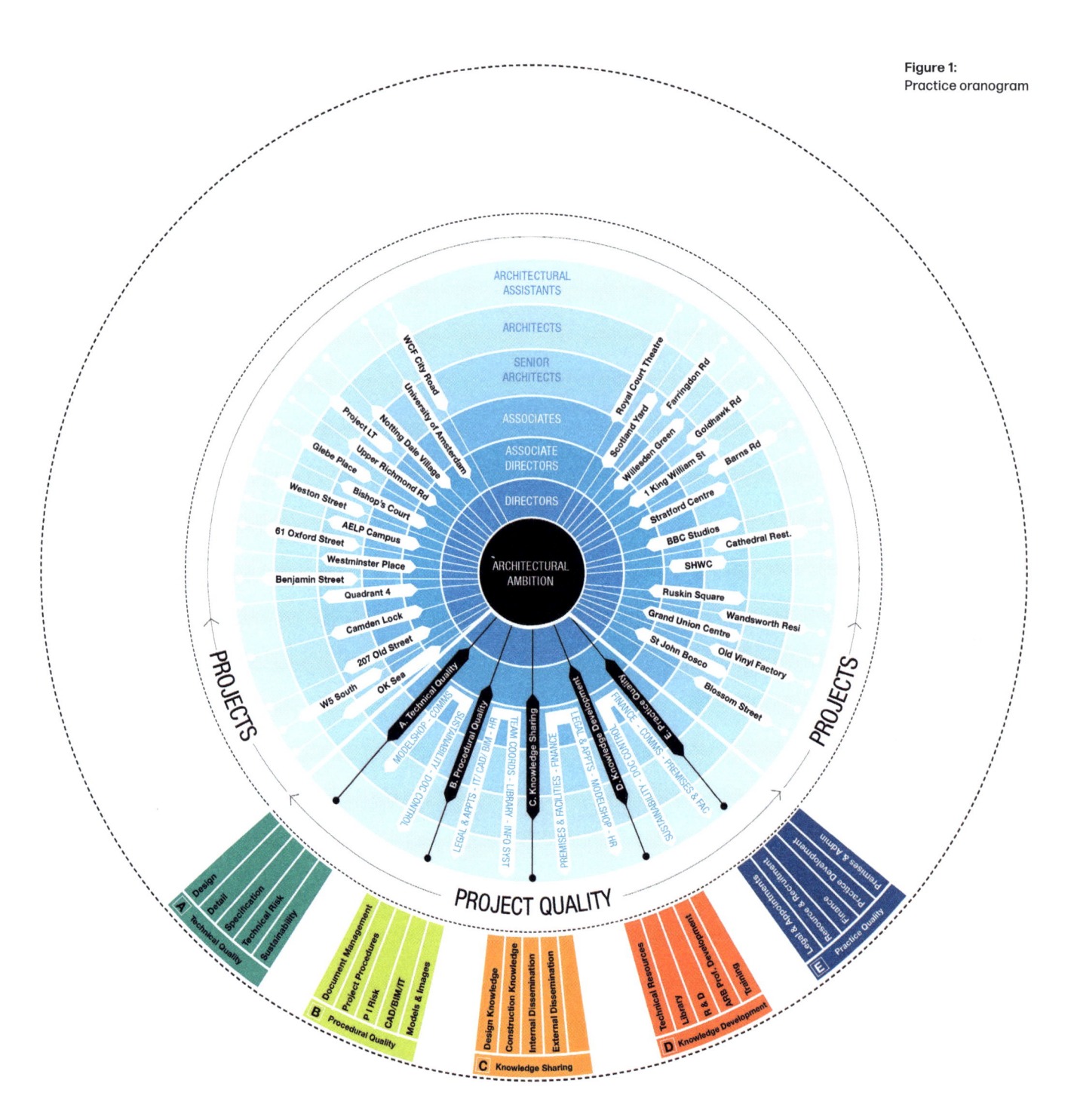

ever better buildings; however, frequent false starts and uncertain future regulatory improvements do not help drive consistent improvement. Regulatory-agnostic performance modelling does.

Designers need to be nimble in their approach, and understand how changing regulations and certification schemes might influence a design and impact clients, and how they might use them as a lever to drive genuine performance improvement in their designs. However, AHMM's focus is on building performance as an absolute: irrespective

84   Energy Modelling in Architecture

of the certification method, regulation or incentive scheme, the practice want to know how their buildings work, how they will perform and how architectural performance can be optimised. Modelling is focused on how their buildings will perform, not on compliance.

Performance is modelled across a range of metrics to support developing designs. AHMM's designers use industry-standard software (IES VE) as well as bespoke parametric tools using Rhino and various environmental plugins like Ladybug and Honeybee. Proposed architectural geometry is interrogated to test comfort, air quality, daylight, glare, management strategies and subsequent energy consumption. This informs decisions about massing, environmental context, materials and wider integration of passive and active environmental technologies.

Considering building performance at the very early stages of the design process enables AHMM's designers to enter into conversations with their clients and design team collaborators with clear performance aspirations and ideas on how buildings will work. Tracking performance through the design and delivery programme helps generate integrated architecture. Focus on performance rather than compliance often uncovers contradictions in these modelling processes. The responsibility of the BPT is to provide insight and information to support architectural decision-making; to justify, explain and develop their thinking; and to optimise and communicate environmental strategies. The challenge is always to communicate design strategies in the most influential way to create positive change.

Process

AHMM use a bespoke 'sustainability toolkit', a design workshop and performance matrix to develop a multifaceted and integrated approach to building performance. Every project is reviewed at key developmental stages: strategy development, integrated design, technical documentation, construction and in-use through a post-occupancy evaluation (POE) programme. The review output is used to highlight opportunities and constraints and develop project strategies, inform specifications and communicate intentions from concept design to post-completion.

In a typical review workshop, capital expenditure, running costs, the human experience of using and making AHMM's buildings, operational and embodied carbon and energy, and wider ecological and human impacts are considered. These parameters and more are set in a formal framework for identifying opportunities, understanding constraints, developing strategies and setting targets.

Project aspirations are represented using a graphically simple 'rose' diagram with a four-point traffic light system and presented back to the team together with a set of notes and items 'for development'. These items may be developed by architectural teams alone, with the AHMM BPT, with the wider design team or through a collaboration between all three.

This approach and capability are particular to AHMM and all new employees are given an induction to the process, the resources available to help with design development and the practice protocols.

The information that the BPT create is focused on influencing architectural decisions. The team work with AHMM's design team collaborators to coordinate performance modelling with any compliance calculations. IES allows for easy sharing of models, theoretical integration with the building information modelling and access to standard weather files, compliance tools and various performance plugins. This information flow aims to develop integrated designs that the whole team, from architect to engineer, and contractor to client, can buy into, understand and deliver. Performance information is visualised and presented to architectural teams and external consultants when appropriate. Visualisations vary from fairly straight-line charts to infographics relating design decisions to the equivalent energy use more familiar to an audience.

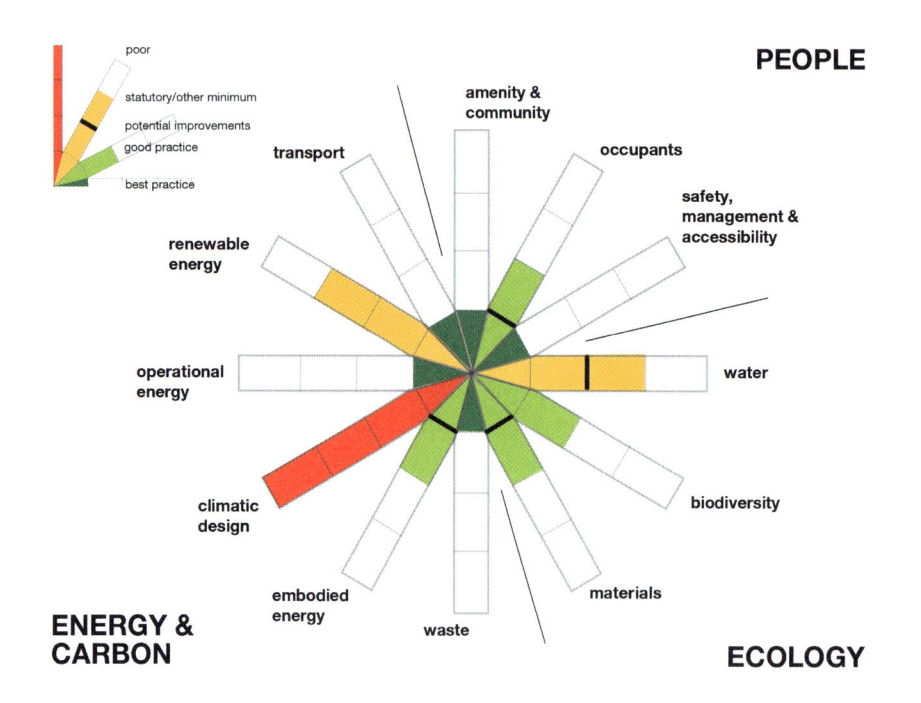

Figure 2:
Rose diagram

Examples

The following examples illustrate the type of energy and performance modelling AHMM's BPT does, and how this is used to influence project outcomes.

Research

AHMM has developed a research programme to complement project work. This has developed in response to multiple architectural teams working on similar problems with common themes. Research is generalised findings from project work or a further development of ideas from projects. It is shared through presentations, information sheets and strategy papers via the practice intranet. AHMM also publish work publicly when possible.

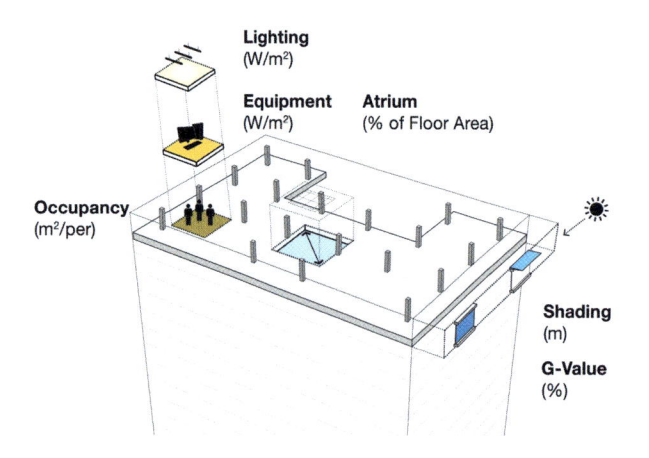

Figure 3:
A concept diagram for a flexible office design

    A paper presented at the CIBSE technical symposium in 2018 explored how a brief for a flexible, speculative commercial office and R&D building could be used to develop a prototype adaptive building for a changing market and climate.

86   Energy Modelling in Architecture

Figure 4:
Opportunity points in the project lifecycle

③ Measure Outputs

Energy Consumption (kWh/m²/yr)
Thermal Comfort (% Occupied Hrs > TM52 Adaptive Comfort) (6)
Embodied vs Operational Carbon (kgCO₂/m²)

Parameter	Replacement
Occupancy	(10 Yrs)
Equipment	(5 Yrs)
Lighting	(15 Yrs)
G-Value	(30 Yrs)
Shading	(30 Yrs)
Atrium	(60 Yrs)
Structure	(60 Yrs)

Timeline: Now, 2025, 2030, 2035, 2040, 2045, **2050**, 2055, 2060, 2065, 2070, 2075, **2080**

Typical Building Lifespan = **60 Yrs**

(Replacement rates over 60 years and opportunities for intervention) (7)

② Apply to Climate Scenarios

Now - London_LWC_DSY1.epw
2050 - London_LWC_DSY1_2050Med50.epw
2080 - London_LWC_DSY1_2080Low90.epw

AHMM used current and future weather files (CIBSE's London_LWC_DSY1.epw for current conditions, London_LWC_DSY1_2050Med50.epw for 2050 and London_LWC_DSY1_2080Low90.epw for 2080) and the architectural and systemic parameters adapted to test performance were occupancy levels, equipment use, glazing ratios, thermal mass in soffits, U-values, airtightness, floor-to-ceiling heights, and heating and cooling set-points.

The purpose of the work was to identify architectural and operational changes that could be made to a building over its life cycle to adapt to climatic changes and ensure ongoing climatic and functional efficiency. The outcomes of this work are informing several other projects currently in development, including a project replacing a 30-year-old economically unsustainable building. The new development must avoid the environmental, material and economic waste of a building with a short lifespan, and AHMM's architecture is explicitly about longevity and flexibility. This short research project is being applied in a real-world project.

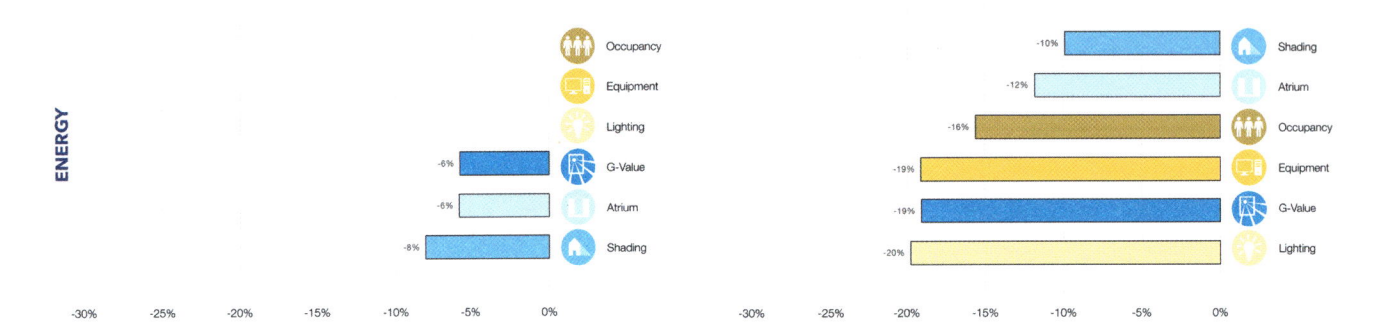

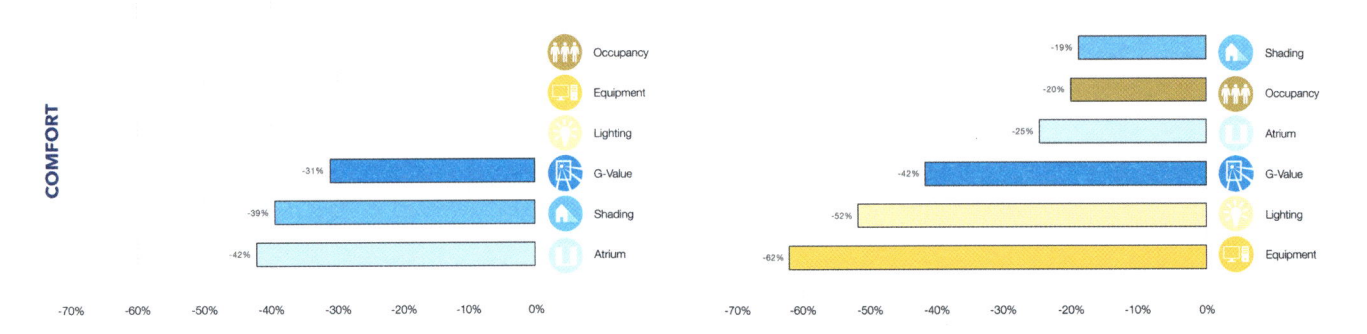

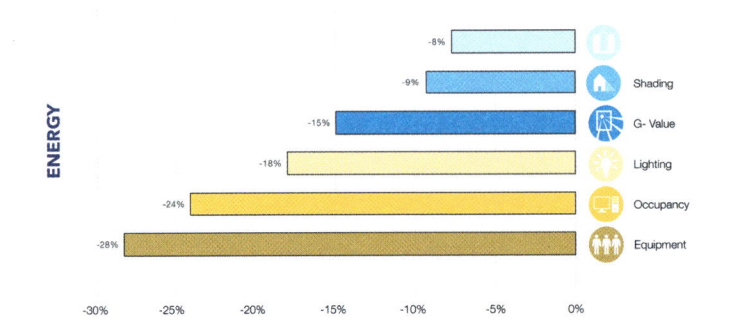

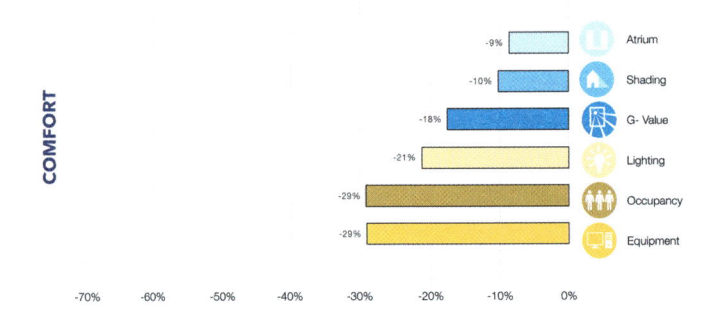

Figure 5: Current and future performance changes through the building life cycle

88   Energy Modelling in Architecture

Challenging briefs

The construction industry has a habit of falling back on known assumptions and technologies, delivering familiar solutions based on buildings that are deemed to have 'worked' in the past. In a recent speculative office project AHMM used energy and comfort modelling to test the briefed and assumed temperature set-points, servicing strategies, management protocols and facade designs.

The brief was based on British Council for Offices (BCO) comfort criteria and a conventional servicing strategy. AHMM's aim was to show that other management and servicing strategies, embedded in an architectural approach, could offer significant benefits in terms of energy consumption, CO_2 emissions, and occupant control and comfort.

The modelling approach developed simple interventions that improve performance, then illustrated more radical changes to further improve performance. In this case the potential reductions in energy consumption associated with a simple measure like changing a set-point were illustrated first, then using opening windows as a proxy for ambient, unconditioned air, AHMM illustrated potential energy-use reductions.

Through opening windows and a night cooling strategy AHMM's team illustrated a 90% potential reduction in cooling loads. The ultimate solution may be a mechanically ventilated, acoustically attenuated delivery of ambient air, but opening windows is a useful proxy. The key point of this work is to open a conversation about different methods of approaching building design, conditioning, and management to provoke change in architectural and servicing strategies.

Figure 6:
Solar gains analysis of a speculative office design

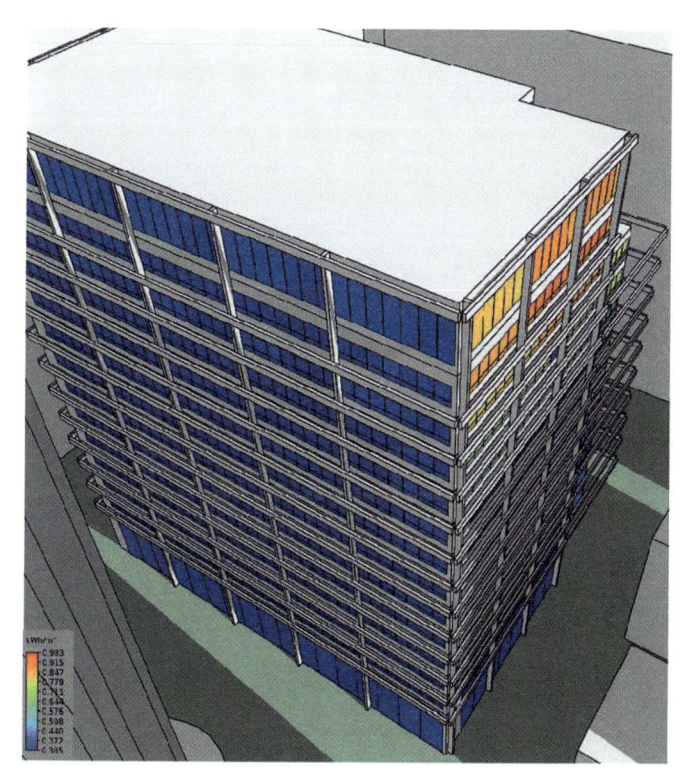

Solar Gains on South Facade 21st June 9am

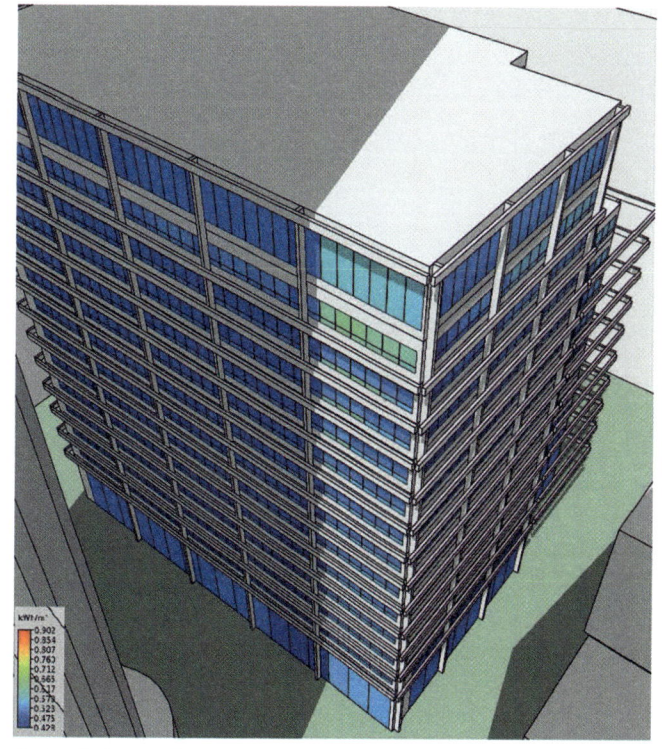

Solar Gains on South Facade 21st June 12pm

Figure 7:
Annual performance improvements

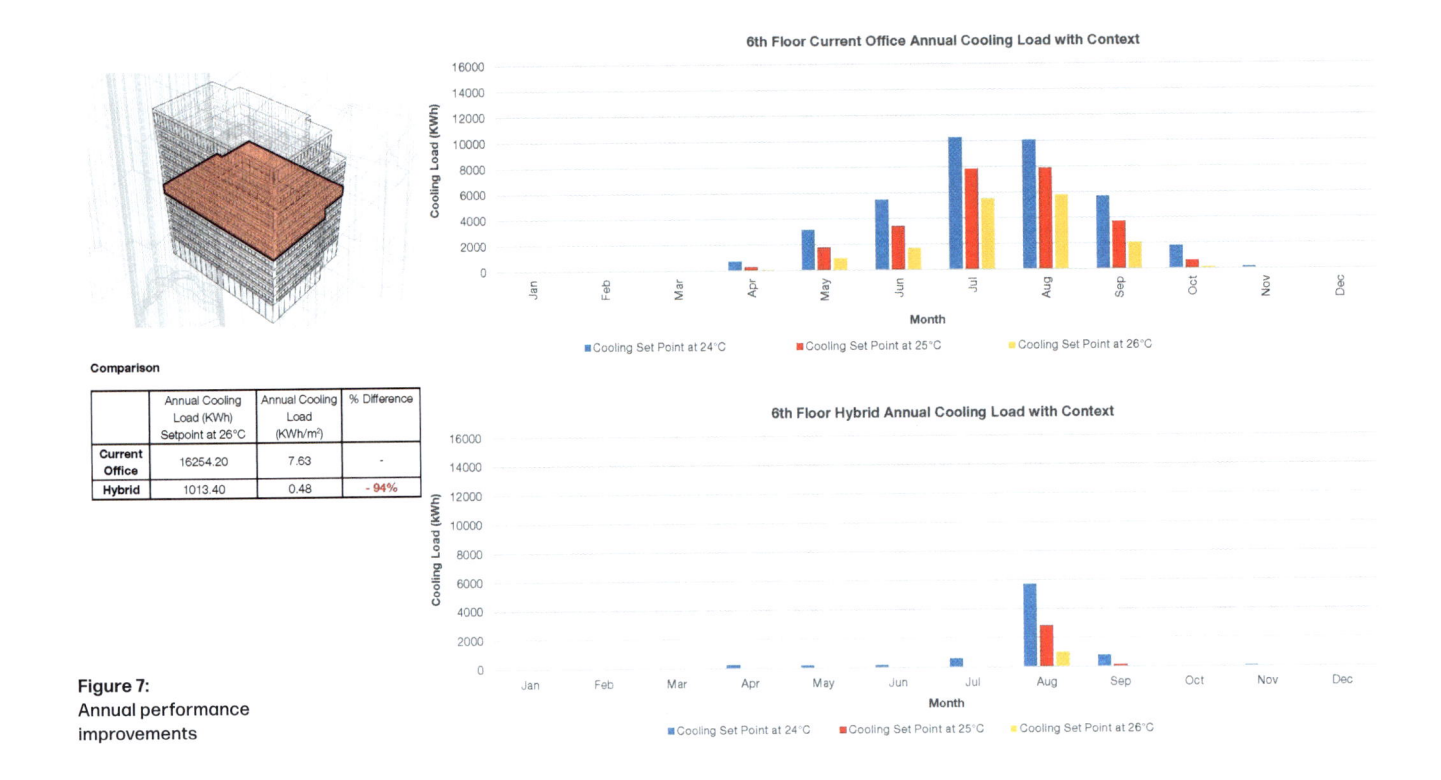

Comparison

	Annual Cooling Load (KWh) Setpoint at 26°C	Annual Cooling Load (KWh/m²)	% Difference
Current Office	16254.20	7.63	-
Hybrid	1013.40	0.48	- 94%

Figure 8:
Daily performance improvements

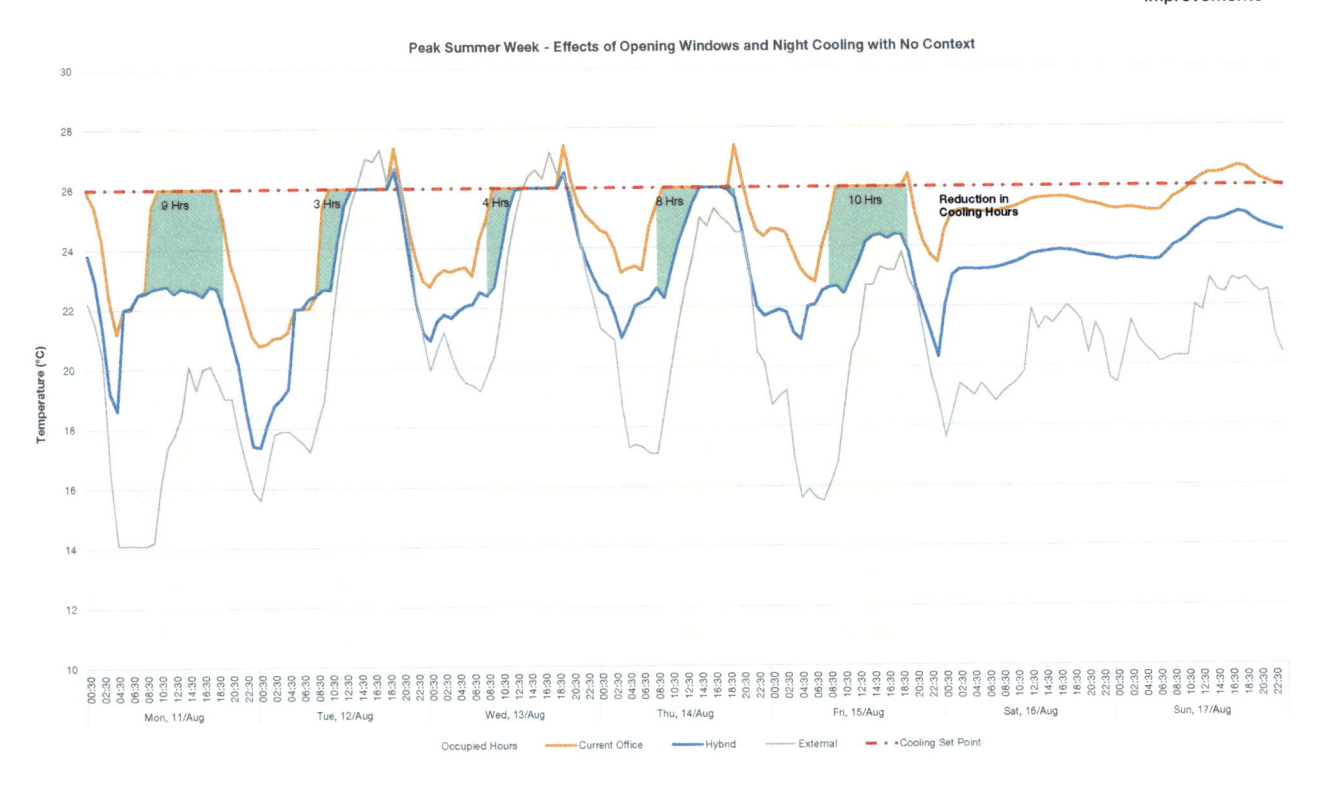

90   Energy Modelling in Architecture

Validating an idea

AHMM's design for a speculative office project in a European capital city challenged a preferred design solution for managing solar gains. This work used a range of metrics to explain and justify an approach to architecture that the design team felt worked on a number of levels: energy, comfort and building user experience.

Figure 9:
Two facade proposals

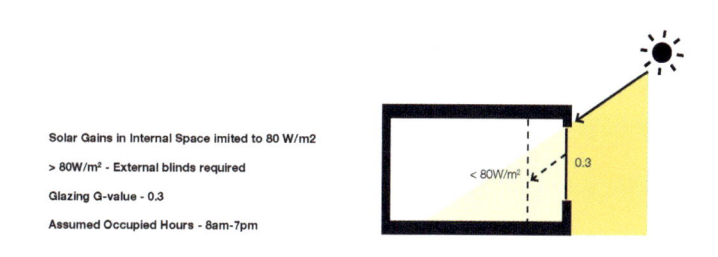

Solar Gains in Internal Space limited to 80 W/m2

> 80W/m² - External blinds required

Glazing G-value - 0.3

Assumed Occupied Hours - 8am-7pm

Plot A Approach - No External Shading

	No. of Annual Hours Solar Gain in Internal Space > 80W/m² and External Blinds Required	% of Annual Occupied Hours Solar Gain in Internal Space > 80W/m²
North East	1	0.02%
South East	838	19%
South West	1307	30%
North West	577	13%

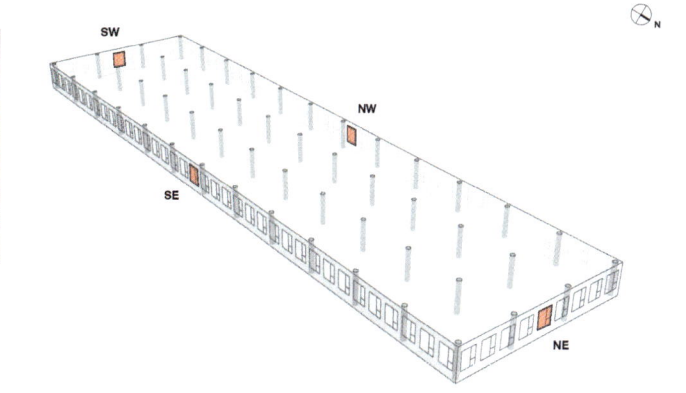

AHMM Approach - Fixed External Shading

	No. of Annual Hours Solar Gain in Internal Space > 80W/m²	% of Annual Occupied Hours Solar Gain in Internal Space > 80W/m²
North East	1	0.02%
South East	1	0.02%
South West	11	0.25%
North West	1	0.02%

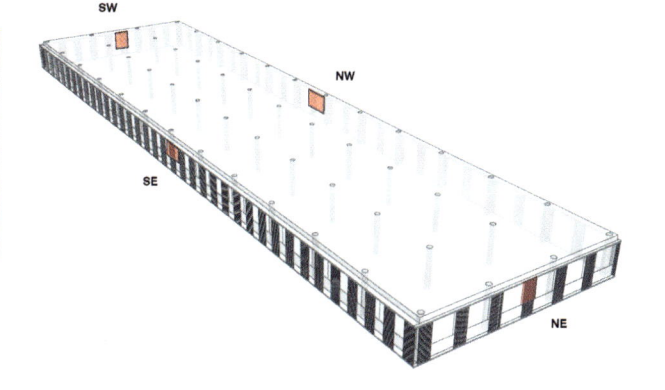

The brief for the project included maintaining views across the city, and the use of a tried-and-tested approach to the envelope design: a solid facade with individual window penetrations and external blinds for solar control. Anticipating maintenance issues with external blinds on a tall building, less dramatic internal spaces and disruption to the desired views across the city, AHMM's proposal was to carefully design fixed external shading which would remove any external moving parts, eliminate all solar gains over the threshold for the proposed servicing strategy and allow for more constant views out of the building.

AHMM calculated the percentage of time when the threshold was exceeded, and how much total time the blinds would need to be down. Illustrating this as an annual chart made the benefits clear and the impact of AHMM's architectural solution fully realised. This shading was integrated into the design team's energy models to verify the outcomes and ensure this was not contradictory to compliance models. Glare and other parameters were also checked.

Figure 10:
Annual charts illustrating time blinds required

Plot A Approach
Total No. of Days Blinds Down - **245**
No. of Days Blinds Down for > 50% of the Day - **148**

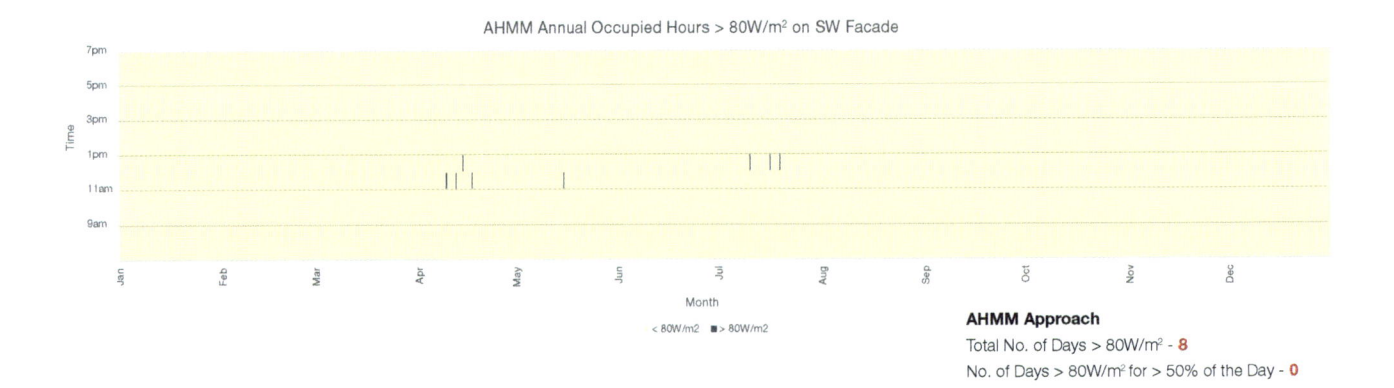

AHMM Approach
Total No. of Days > 80W/m² - **8**
No. of Days > 80W/m² for > 50% of the Day - **0**

An architectural idea is the start of the process; performance modelling refines and verifies that approach. Communicating the results and opportunities is often the difficult part. This project illustrates the need for clear and compelling graphic communication of modelled outputs.

Design integration
Incorporating environmental strategies into architectural proposals, particularly in commercially driven projects, can be challenging. Modelling therefore often focuses on not only illustrating how a designer's proposals will perform, but how they are driven by the architectural treatments and can meet market requirements.

With an architectural ambition to expose concrete soffits, this work sought to illustrate the potential to reduce the perceived need for comfort cooling in residential apartments and reduce waste and applied finishes.

Figure 11:
Position of thermal mass

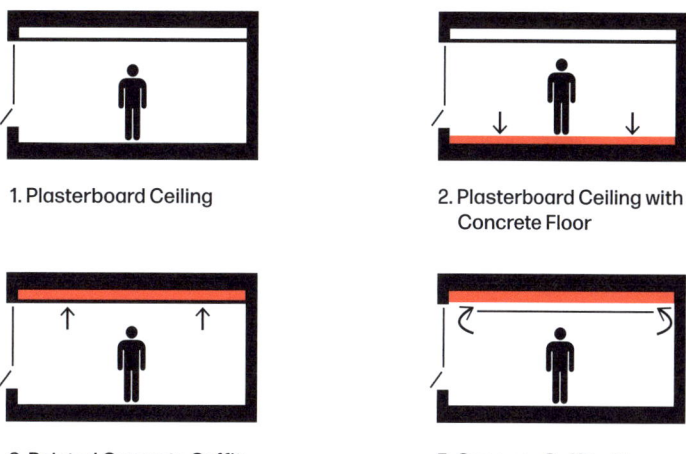

1. Plasterboard Ceiling

2. Plasterboard Ceiling with Concrete Floor

3. Painted Concrete Soffit

5. Concrete Soffit with Plasterboard Raft

There are advantages to be exploited in the use of thermally massive elements in residential buildings with effective 'domestic management': maximum temperatures can be reduced, often completely mitigating the perceived need for comfort cooling. It is possible to illustrate through user guides how seldom active cooling is required; user engagement is increased; connection to the changing seasons is enhanced; winter temperatures are increased; and applied finishes and the resultant waste is reduced.

Figure 12:
Performance profiles of thermal mass

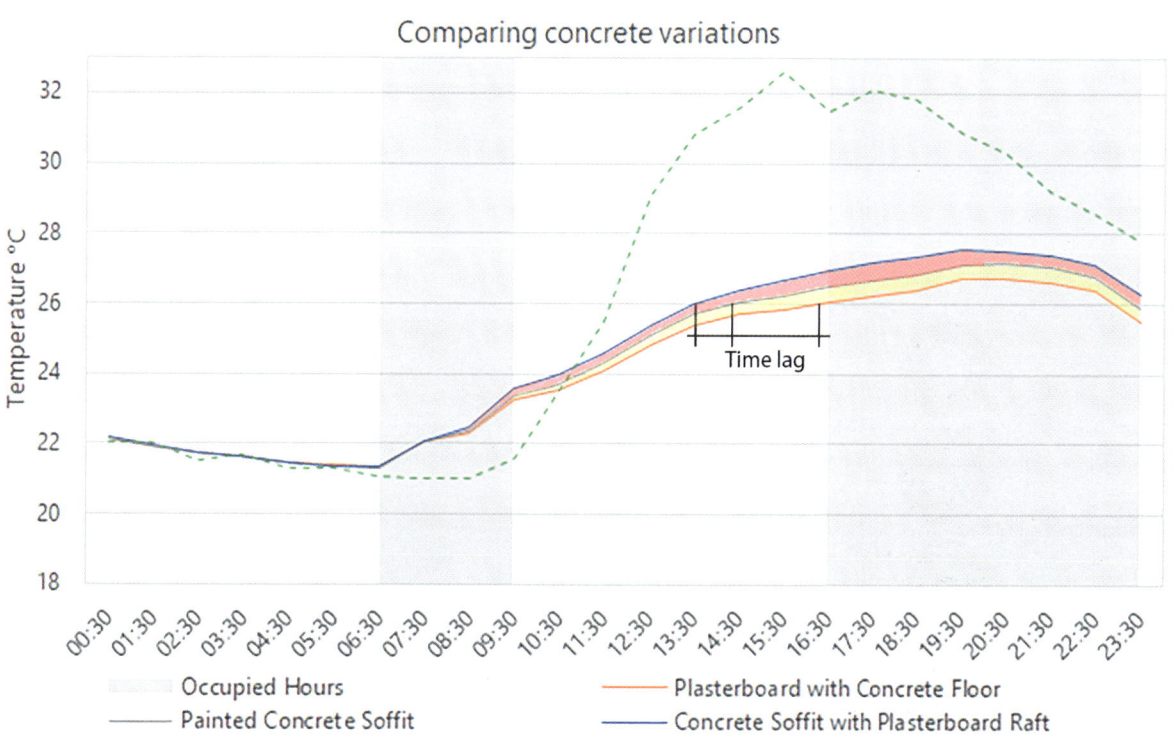

AHMM – from programme to practice

The work showed that on the soffit, the design can reduce summer temperatures by around 2°C and increase winter temperatures by around 1.5°C. If the design uses concrete in areas with less visibility such as floors or behind ceiling 'rafts', or it is simply painted white, similar benefits apply and can add value to the project. In this project AHMM deliver an architecture that includes environmental benefits and they are careful to illustrate the benefits, particularly where this is a perceived challenge to the market.

Compliance

While AHMM's focus is always on performance in use, this can contradict the modelling processes required for compliance with Building Regulations. Performance modelling must illustrate how an architectural approach to building performance will not only improve performance in use, but how it will help (or at least not hinder) the path to regulatory compliance. This is especially true in overheating analysis, energy and Part L calculations.

Scenario 3
50% shading
Night cooling

50% solar shading.

Windows open at 21° internal temperature, closed at 25° external temperature betwen 07:00 and 22:00.

Fabric night cooled between 22:00 and 07:00.

Up to five people in the house, variable occupation throughout the day.

Figure 13:
Occupancy schedules

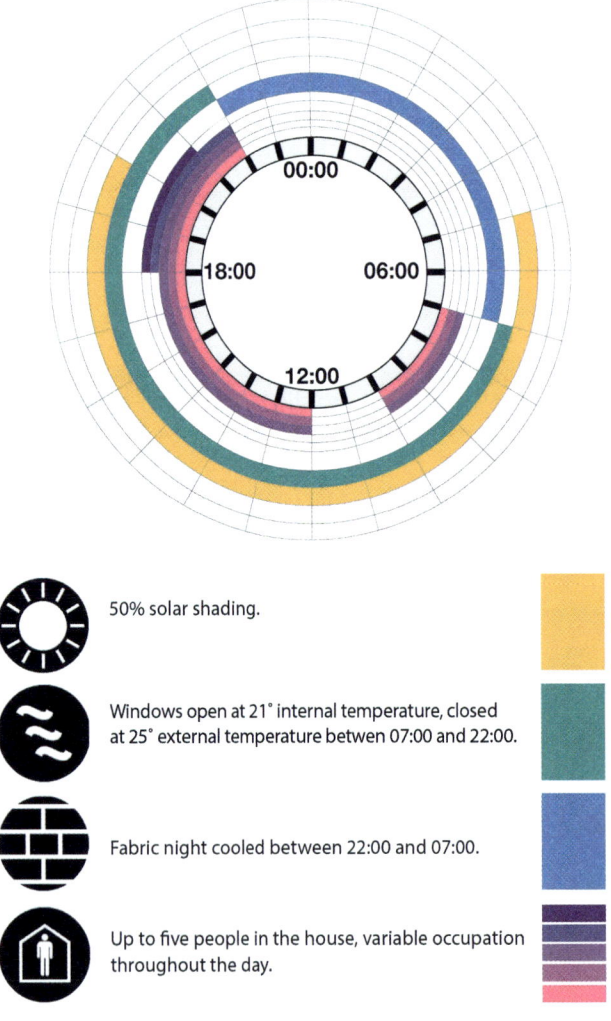

94   Energy Modelling in Architecture

For example, when carrying out comfort studies AHMM always use CIBSE TM52 as a reference point, and present temperature data in graphic formats for lay audiences when required. For domestic clients the practice set out temperatures throughout the day based on their expectations and analyse the project based on the client's anticipated use patterns. The design team can then communicate with the client on their terms, but also apply TM52 to gain cross-project comparative insights and a measure of compliance. The team also check against TM59, but only as a compliance measure. TM52 gives the flexibility to test against AHMM's own and the client's parameters.

Figure 14:
Example performance

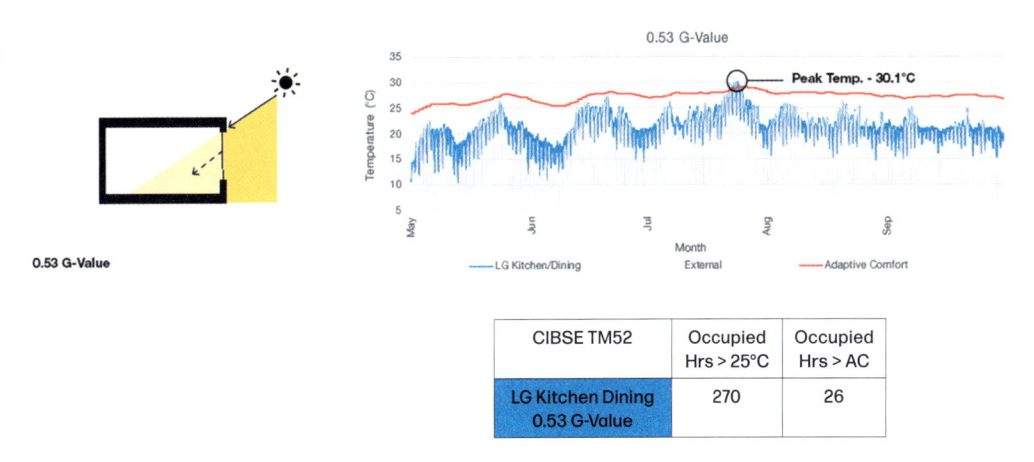

CIBSE TM52	Occupied Hrs > 25°C	Occupied Hrs > AC
LG Kitchen Dining 0.53 G-Value	270	26

For energy, AHMM use heating and cooling, regulated energy uses and occupancy schedules that are compatible with SBEM calculations to ensure the design development is compliant, but run a separate performance model to test the architectural designs against more realistic dynamic operational strategies and performance models. The outputs are designed to reassure designers that proposals function properly and to challenge the design team collaborators to be ambitious rather than to deliver mere compliance-driven solutions.

The value of in-house expertise – conclusions

Building performance modelling is discussed above as an integrated part of AHMM's architectural processes, providing potential for robust analysis during early design development. The use of early building performance modelling presented designers with informed choices to develop ideas that have their genesis in projects and in independent research projects. With robust analytical capacity, the practice suggest that the design team can challenge clients' expectations of what a building will be and how the practice will deliver their aspirations through architecture, management and technology.

As the urgency with which we need to tackle climate change increases, architectural practices' processes, metrics and ambitions must become more robust and our outcomes more ambitious. The industry must develop protocols to deliver, test and validate environmental strategies.

Chapter 9: Feilden Clegg Bradley Studios – the individual and the team

Feilden Clegg Bradley (FCB) Studios' view on environmental design was largely forged through the thinking reflected in Peter Clegg's book *Energy for the Home*,[17] setting out principles for low-impact methods of powering the home. Despite being published over 40 years ago, many of these underlying concepts remain valid today, such as passive solar design, on-site power generation and water recycling.

Figure 1:
Energy for the Home

FCB was at the time of the book's publication a small, growing practice, which enabled the book's underlying principles to permeate through the projects. There were fewer people and projects to oversee, and sharing resources between projects was straightforward. As a small practice, there was also a more unified culture, with the people drawn to it at the time having a stronger alignment with the practice principles, notably sustainability. This led to more autonomous sustainability within the projects; each person brought the principles of the practice to the projects because they were personally aligned with them, rather than acquiring them through training and oversight as would be necessary in a larger practice. There was also reduced separation between people and the range of projects, enabling greater informal sharing of information; a simple chat over a coffee can cover the entire project history of the practice, enabling sharing of important lessons between people and projects. Each project in effect received a form of passive supervision.

In the 40 years of FCBStudios, the practice has grown to over 200 people, across four offices, split into ten groups. As the practice grew, so did the breadth of knowledge, from building sectors to design philosophies, all playing a part in making not only a diverse practice, but also a more resilient practice that can absorb market changes in the construction sector. This feeds into FCBStudios' collaborative design philosophy, creating buildings that closely fit the needs of both the clients and the end users, rather than clients buying into a specific FCBStudios design/style.

However, the downside to this expanding and broadening of the practice is that the implicit passive supervision that worked before becomes less attainable; knowledge is no longer held collectively and instead risks becoming siloed within offices and groups of people. This has a significant impact on all knowledge-sharing, but particularly on ensuring that the best practice of sustainable design is shared between projects. It is no longer the case that an incidental chat over a cup of coffee will transfer knowledge across the practice; there will always be people in the other offices who cannot join in and bring their own experiences.

As a consequence, this has increased the importance of active supervision, where each project receives a formal review at key design stages (typically aligned with the RIBA work stages). At FCBStudios, these reviews have typically focused on the design and contractual aspects, but with an ambition to continually improve the environmental and social performance the practice now has dedicated sustainability reviews. These sustainability reviews take a holistic view of the project, firstly getting an overview of targets set within the brief, but then moving on to the broader sustainability aspects, using the One Planet Living framework for guidance.[18]

Figure 2:
One Planet Living framework

- Health and happiness
- Equity and local economy
- Culture and community
- Land and nature
- Sustainable water
- Local and sustainable food
- Travel and transport
- Materials and products
- Zero waste
- Zero carbon energy

As a One Planet practice, these principles permeate through FCBStudios' decision-making, creating a culture that takes a broad view on sustainability, from more typical environmental aspects of the practice's operation, such as reducing paper consumption or improving energy usage, to making the practice vegetarian or promoting apprenticeships as an alternative low-cost route into architecture. These One Planet principles are championed by eight members of staff who have trained as One Planet representatives. They cover a mix of roles and provide oversight of the whole practice, acting as resources and influencers in practice decisions. This is supported by the larger research and innovation group, which acts as a forum for sharing knowledge across the offices and groups within the practice.

Within the sustainability reviews, FCBStudios consider how each of the One Planet principles is addressed through the design process to improve performance in construction and operation. Standardised actions are proposed for each sector the practice works within, such as housing. These are intended to showcase the types of interventions the project could make, acting as not only a way of recording the actions made, but a prompt for a conversation around the project and the One Planet principles. This is an emergent list, based on best practice from the studio's experience and the wider industry, but as more projects engage with the reviews, the list increases, effectively sharing the knowledge across the practice during the design process. These interventions and actions are weighted according to their impact and how pioneering they are, to give an idea of the relative performance of projects within the practice. These weightings are reviewed by the One Planet integrators to ensure broad compatibility, but they are not the focus of this exercise.

The One Planet Living framework is a useful communication structure through which every project's unique set of constraints and opportunities can be reviewed. The dialogue and awareness of the principles are key to informing the outcomes. Commonly, these conversations give rise to ways to push the project forward on broad sustainability principles such as engaging with the community, or increasing tree canopy cover, on others leading to very specific goals that align with client concerns. This is as much about providing the design team with the framework for thinking about sustainability as it is about reviewing the project. By empowering the architects with the knowledge and expertise to discuss sustainability knowledgeably, they are able to act as sustainability ambassadors within their projects, the practice and the wider community.

Environmental modelling within the practice

As an architectural practice, FCBStudios are often working at the inception of a project, when the greatest impacts on environmental performance can be made. At these early stages the practice's main focus is maximising the passive design principles, setting a solid foundation for the environmental strategy as the other consultants become involved. However, the design team only use a limited suite of tools during the initial phases of the project, focusing on form, orientation, daylighting and building fabric. This reflects the low level of information available at this stage; buildings tend to be simple massing blocks rather than detailed designs. Once past the initial stage of design, typically up to the end of RIBA Stage 1, the designers expect the consultant team to provide more detailed modelling and design feedback, particularly around the energy performance.

On dense sites, a key step in the design process is understanding the daylighting in and around the building, getting an idea of overshadowing from neighbouring buildings, and the impact of the proposed building(s). Much of the modelling at this early stage is undertaken using SketchUp. When the site context is modelled and geolocated in parallel with the proposed buildings, the design team can very quickly get an idea of how the sunlight moves through the site using SketchUp's built-in daylight function. This is key for understanding general constraints, as well as sunlight and sky visibility constraints on the site. The Building Research Establishment's (BRE) site layout planning for daylight and sunlight sets out guidelines for access to direct daylight in public-realm areas, and is often quoted by planning authorities as a requirement to be met.

Overlooking this at the start of the project could instantly undermine the project's viability, and so must be considered. SketchUp is not natively able to show the annual summary of daylight availability around the building, so designers at FCBStudios use the Sefaira plugin to assist them. They have found that there are few programs able to show this measure clearly, with a majority focusing on internal light levels. Using Sefaira as a SketchUp plugin fits the design team's typical workflow, with the existing 3D models able to be used with little modification required. The proposed designs can then be iteratively modelled to ensure they are creating outside areas that will be usable throughout the year.

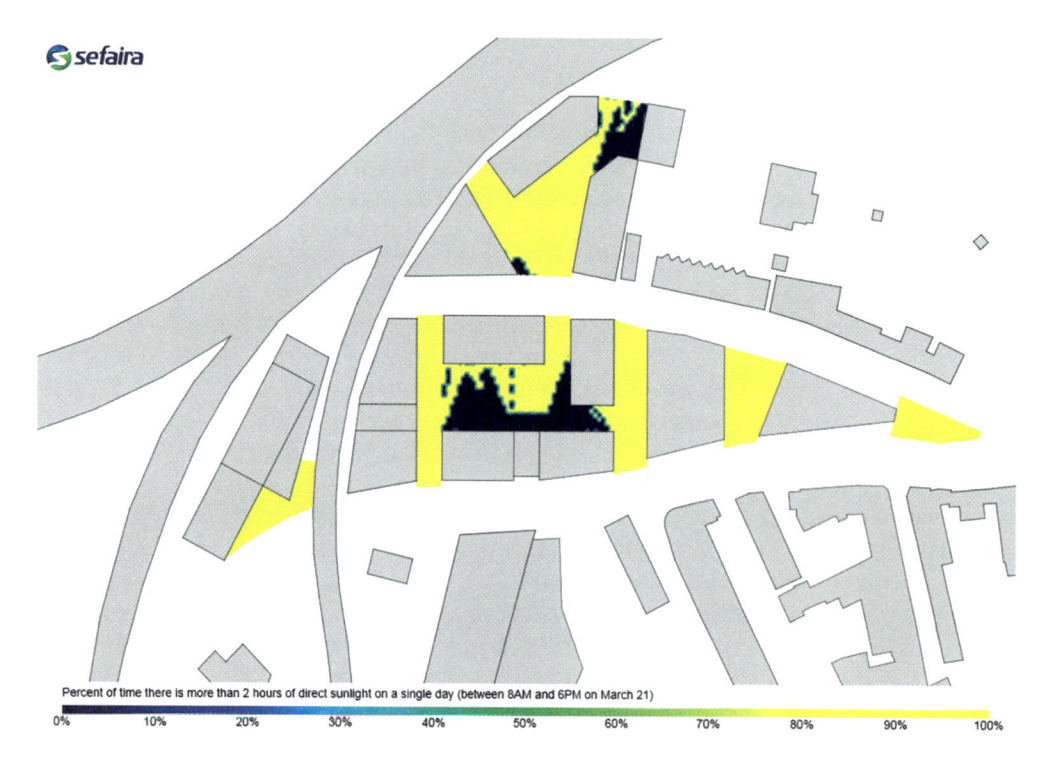

Figure 3:
Output from Sefaira software showing light availability in external areas

Alongside modelling the external daylight availability using Sefaira and SketchUp, the design team assess the internal daylight availability within the proposed building. Their aim is to create spaces that can be fully daylit throughout the year, reducing the need for artificial lighting and the associated energy use. Advanced modelling techniques such as climate-based daylight modelling are incredibly useful indicators for understanding daylight autonomy, but with so many unknown variables at the beginning of a project, any results can be misleading. Instead, they rely on the simple daylight factor, aiming for between 2% and 5% to ensure daylight autonomy, based on the old Building Bulletin 90 guide for school daylighting. FCBStudios describe this as a 'sweet spot', with more than 5% significantly increasing the chance of overheating due to solar gain, particularly on west-facing facades. As the design progresses, this overheating risk is fully assessed by FCBStudios' engineers, where they can act to reduce the amount of daylight, but it is often difficult to increase the amount of daylight as the design progresses.

To analyse the daylighting within the buildings, the team select key floors within the model, particularly lower floors that are at risk of overshadowing and upper floors that are at risk of overheating. They have tried many software packages to understand the daylighting, but the simplest workflow has been found using the VELUX Daylight Visualizer. This program has the ability to import SketchUp models, and has a quick-rendering engine, enabling a relatively fast and simple iterative design process. The simple graphics mean the daylighting strategy can be quickly and simply conveyed to a non-technical team.

The key to this early-stage daylight modelling is that it enables the design to progress until the design team is expanded to include specialists who can provide robust advice. Much of the modelling in these cases is confined to within FCBStudios, acting as internal checks on their design. However, where projects require and enable an enhanced modelling process, the practice have developed bespoke modelling processes.
At Broadcasting Place for Leeds Metropolitan University, the facade allowed for a pseudo-random pattern, enabling a generative computer algorithm to be used. For this development, the facades were modelled in Excel, with each cell representing a 1.5m-wide module per floor. Previous modelling by the BRE had provided figures for the access to daylight within each module, which was entered as a percentage. Using VBA scripting, these access-to-daylight values were analysed against targets for solar gains (set at 20W/m^2), generating a permissible glazing percentage and glazing type (solar control or normal) for that module. The script created a range of options that was then selected for aesthetics, ensuring the ultimate design control resided with the project architects. While Excel seems like an unusual choice for designing a facade, it provided the ability to use existing data from the BRE and could leverage the scripting of VBA, but also feed directly into the MicroStation CAD files to produce the facade pattern.

Much of FCBStudios' in-house modelling capabilities revolves around daylighting, but through taking inputs from external engineers, such as the daylight availability at Broadcasting Place, they are able use the engineers' environmental data to underpin their design. For the Hive library for the University of Worcester, the unique roof shape was crafted to maximise the daylight and wind-driven ventilation, all while working within the parameters of buildability for a cross-laminated timber (CLT) roof. In-house, FCBStudios do not have the ability to analyse the wind-driven ventilation, but their project engineers, Max Fordham, were able to provide data on the wind movement around the roof lights, setting out design rules to ensure optimum ventilation performance. Using these ventilation parameters, along with the limits of the construction method developed by the contractor, FCBStudios have created a script within GenerativeComponents for MicroStation to create the roof elements. This allowed the architects to manipulate the roof design quickly to achieve their desired aesthetic, but without hindering the environmental and construction parameters.

Working with the design team to create constraints gives designers a freedom to model within those parameters, ensuring that they each work to the team's strengths, but most importantly remain within their scope of services. FCBStudios' use of modelling on projects is predominantly inward-focused, acting as guide for their design, and is rarely conveyed to the client team. This is mainly to ensure that they do not take on liability beyond their services, which becomes particularly complex as the design progresses and it overlaps with the work of other consultants. Ensuring clarity of design liability is a key issue around environmental design in buildings; it straddles all aspects, but the ultimate responsibility needs to remain with the correct parties, usually the engineers

100% Glazed   75% Glazed   50% Glazed   25% Glazed

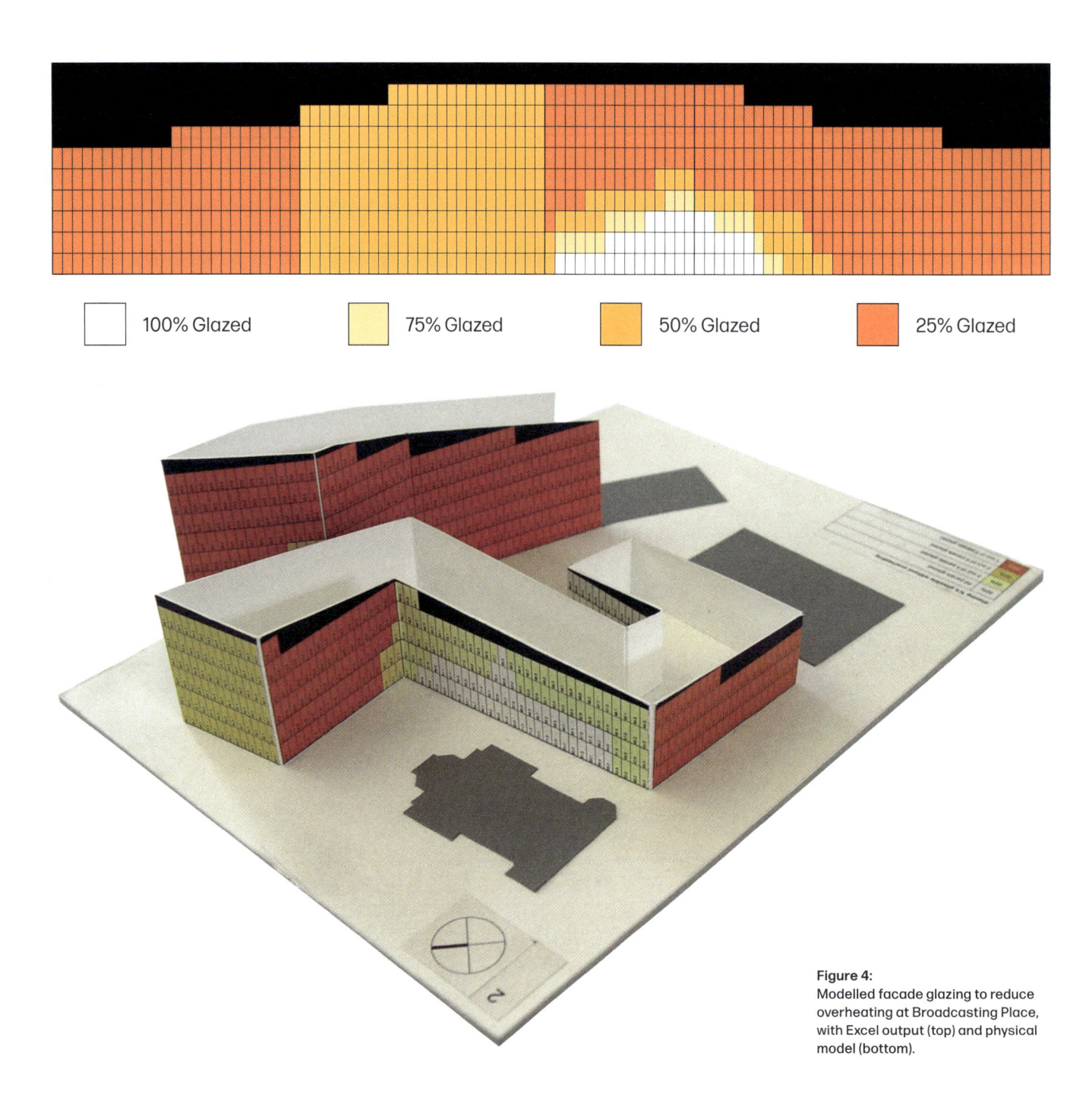

Figure 4:
Modelled facade glazing to reduce overheating at Broadcasting Place, with Excel output (top) and physical model (bottom).

Figure 5:
Cross-section through the Hive building at the University of Worcester, showing the optimised timber roof cones and the spaces they create underneath

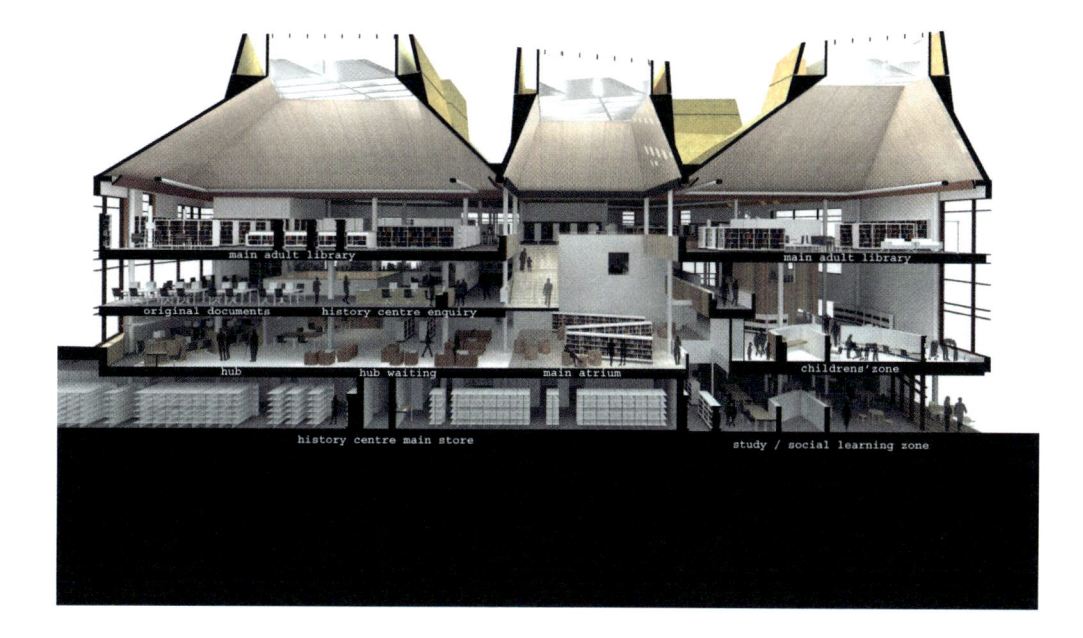

or the environmental consultants. Through the design team setting constraints for the FCBStudios team to work within, the liability for the elements of the design (such as the ventilation) remains with those who have set the constraints, freeing up the design team to work within those boundaries, whether using generative design algorithms or more labour-intensive iterative design processes.

Creating rules of thumb

Detailed modelling of the building energy performance, whether the whole building or individual elements, remains out of scope for the project teams within FCBStudios. Instead, embedding the sustainable principles into the projects relies on broad rules of thumb that create a starting point for the design. The aim of these simple principles is to set up the conditions that enable a sustainable building to be produced, embedding them from the very start of a project. This ensures that the building does not exclude any opportunities for reducing the energy use, because as the project progresses significant changes become more difficult and carry an increasing cost. Getting these concepts in early ensures that the whole project team is aware of the sustainable principles from the outset, but can also act as inspiration for the team, setting the benchmark for expectation.

The key to these rules of thumb is that they must be underpinned by robust research, typically requiring some form of modelling using specialist software. Often, this software is not available to the design team, particularly expensive full-dynamic simulation software, but also it commonly requires specialist skills. Instead, FCBStudios' rules of thumb are the subject of a focused piece of research, testing the impact of a building element across a range of scenarios. This work is led internally, but where specialist modelling/knowledge is required that cannot be internally resourced, they reach out to manufacturers, other consultancies and universities to provide the expertise. In all cases, the aim is to embed the knowledge in a simple design rule that the project teams can follow without heading back to first principles. As the design progresses, the team can rely on the rest of the design team to help with additional modelling to understand the performance in each specific project, but these rules of thumb provide the best 'first guess'. Many of these were formalised and published in the *Environmental Handbook*,[19] enabling the wider industry to benefit from these simple guides.

A recent example is the default external wall thicknesses FCBStudios use when calculating the indicative gross external area (GEA) to gross internal area (GIA) ratios. These ratios are often used to indicate the efficiency of a development, with a temptation to squeeze the wall thicknesses to increase the efficiency on paper. However, without the in-house ability to accurately model U-values to indicate thermal efficiency of the walls, teams cannot be sure of the impact of this. Instead, a focused piece of research created a set of indicative wall build-ups using a number of facade materials and insulation types. These were modelled by the manufacturers of the insulation to give a good idea of what wall thicknesses were required as a starting point. FCBStudios expect projects to begin discussions on the GEA/GIA ratio to include walls that are 550mm thick as a minimum. This will achieve a U-value of 0.13 using most facade materials, but also ensures that there is some flexibility for unusual details within the wall build-up that will affect the thermal performance (such as columns). By the time the project begins to look at detailed design, the designers can rely on the other consultants and manufacturers to provide increasingly project-specific advice.

Conclusions

Within FCBStudios, much of the designers' work is backed by processes and research that has used detailed energy modelling. The findings from the energy modelling, whether generic for broad application, or project specific, provide the design team with clear findings for them to make informed decisions. It is at the early stages of design where they can have the biggest impact, getting the basics in place to make a low-energy building. This is where understanding principles such as daylighting and building fabric can have the most influence, and their limited use of modelling expands. As their designs move through the RIBA work stages, they work with the wider design team, using their expertise to guide the designers through rather than develop their own in-house modelling team. This keeps the division of responsibility clearer and increases buy-in from the whole of the design team.

As the practice has grown, they have learned that they need greater oversight and guidance to create truly sustainable buildings. Sustainability is not something that just happens, it is crafted and needs support to flourish. FCBStudios dedicate time on each project to consider the sustainability opportunities, using this period of reflection to look at best practice within other practice projects. This is not a simple one-hour review but the start of the discussion, creating connections between projects and people to share knowledge. These personal connections are not only incredibly effective at spreading knowledge, but also create lasting bonds between offices and groups.

In parallel to the sustainability reviews, a key aspect of the success of the practice is in creating an atmosphere of sustainability. Using the One Planet Living framework not just for their projects but also to inform how they operate as a practice ensures it remains at the forefront of each staff member's minds. As they are discussing the relative sustainability of different fruit-supplier business models for the practice, they are also raising awareness of the business practices for suppliers/subconsultants/subcontractors for their projects. Knowing that there is a question to be asked opens the way to true sustainability.

Chapter 10: HOK – achieving contextual design through a measured process

The importance of weaving an analytical approach through a creative design process was advocated by Louis Kahn, who called for a greater integration of the unmeasurable and measurable in the building design process. The built environment cannot and should not be designed without addressing its physical, social, cultural, economic and environmental context. Kahn's call for an integrated ethos can be applied quite literally to the environmental context, where designers can truly 'measure' or predict a host of metrics. At a micro scale, they can use specific analytical predictions such as daylight and glare to enhance occupant comfort and wellbeing, and at a macro scale the same strategies can help reduce energy use by reducing globally harmful GHG emissions.

Defining and designing a process to thoroughly integrate environmental analytics into design on paper is one thing – but it is a completely different effort when it comes to actually implementing it in a large design practice such as HOK. The firm has a global staff of over 1,600 with roughly 24 offices around the globe – which translates to a large number and variety of both designers and projects. A firm-wide structure of sustainable leaders and specialists work with project teams and provide necessary support.

Figure 1:
Structure of sustainable leadership

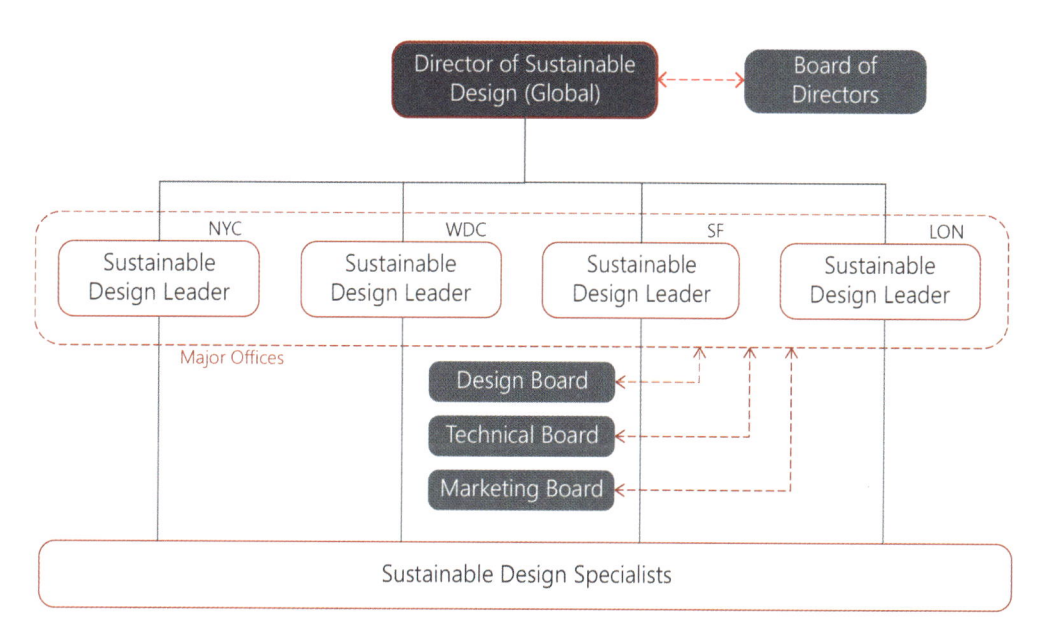

All the major offices have a designated sustainable design leader (SDL) who orchestrates sustainability integration in all projects within that office. This leadership group may have expertise in various aspects of sustainable design, but their primary task is to provide inspiration, define sustainability goals in coordination with global teams, and lead their studios towards achieving such goals.

As part of this role, the SDL engages with designers early on individual projects to define sustainability goals and assist the team in ensuring that appropriate environmental analyses are performed before a design is finalised. Although each studio has its own processes in place, most SDLs will be part of project initiation meetings and even project pursuits to ensure sustainable design goals are discussed and a sustainability framework is established to achieve such goals. It is pertinent that the entire design team (internal and external) are collectively aware and agreeable to a framework early on in a project. The specifics of the processes and methodologies of environmental and energy analysis can then follow with ease.

Overarching performance goals

It is important to establish collective goals for a design firm upfront, both in terms of process and convictions as well as desired outcomes. The latter is often easier to accomplish, especially with the help of existing benchmarking tools available such as AIA's 2030 Commitment, to which many firms have signed up. AIA's 2030 Commitment has become a widely accepted platform in the United States, created by the American Institute of Architects (AIA) for architectural firms to collect and submit performance data for individual projects and continuously track their progress towards a goal of achieving net-zero-energy buildings by 2030. Establishing clear collective goals, on the other hand, requires an organisation to reflect on its culture and determine its core design ideology – a design manifesto perhaps – that in turn helps create specific guidelines to achieve a desired outcome. A studio's own design philosophy needs to embrace, adopt and own a sustainability manifesto that is inseparable from its fundamental principles. Such an approach will ensure that sustainable design in the practice is not seen as an 'add-on', but rather an integral part of the design philosophy of the office.

Establishing guidelines and processes

Apart from administrative goals to encourage all staff to achieve certifications, ensuring most, if not all, projects are targeting third-party certifications (such as LEED), a firm-wide decision at HOK has been to mandate that all large-scale projects must engage with sustainability specialists and at the very least run an energy model, regardless of whether it was required by the client. For projects with interior design scope, the lighting designer is tasked with providing a final lighting power density (LPD) analysis for all projects. The intent of these mandates is essentially to reinforce the firm's commitment to all project managers in all studios, so they may ensure that project schedules, contracts and fees aptly reflect and accommodate required resources for energy and environmental analytics. Each project can then determine whether specific tasks for the project are performed by specialists in-house or outsourced to specialist consultants. These mandates are also in alignment with AIA 2030 Commitment submissions, which makes it easier for the entire firm to collate project performance data, submit to AIA and monitor their progress over successive years as well as against the wider (anonymous) industry data. For a large firm such as HOK, with numerous global offices, AIA's platform has helped track the firm's successes and failures and even compare performances between all the offices. Such commitments also motivate the leadership, designers and project managers to ensure that appropriate analysis is being performed for all projects.

In addition, the sustainability teams collectively across the firm have defined clear processes that guide design teams with the types of studies and analysis that must be performed during the design development of a project (see Figure 2). This process is important given the fast pace and other demands for project deliverables where performance analysis could easily be overlooked. It is rare that an initiative such as developing a global process map is undertaken by leaders from different offices and then shared with the larger group for their review and feedback. Although the process map is shared by all, different studios and teams may modify or use portions of it depending on specific project needs. Also, since this design process is aligned with a general design development process, it has universal applicability even in differing geographic locations and varying local compliance requirements. Providing guidelines and technical support along with mandating minimal measurement metrics for projects is a general approach by the firm to reach its self-defined goals for high-performance design.

Figure 2:
HOK's 6 Steps

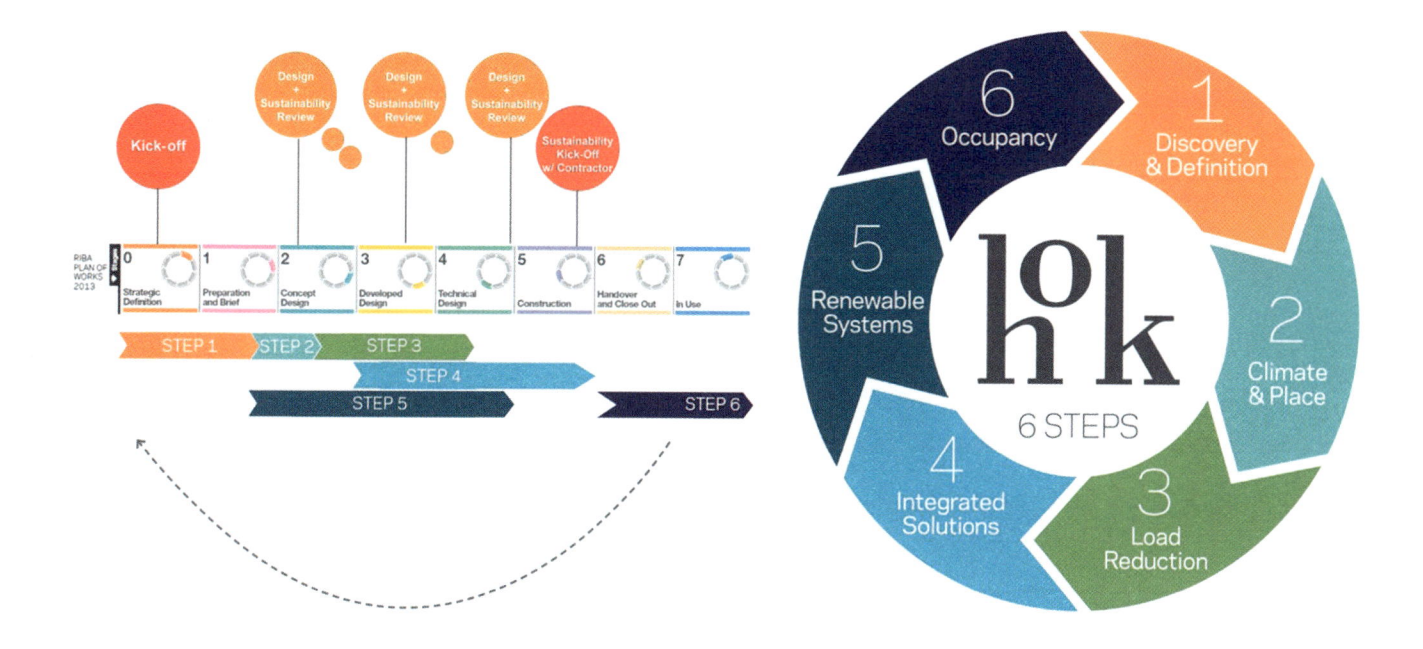

Leading up to predicted energy use

It has been established that using environmental analytics including, but not limited to, energy models early in the design process is the correct approach for sustainability integration. Although firm-wide, many different environmental analysis tools are used by HOK's design teams, a general lean has been towards analysis tools that are easily integrated with design software. The information derived from such analysis should be re-applied to the design process for the most benefit. It should be added that such analysis can be done incrementally and sometimes independently to evaluate specific metrics such as daylight, glare and thermal comfort – although ultimately such studies will collectively lead the design to achieve energy efficiency. Often, very specific environmental analyses are more beneficial to the design team in making the right design decisions. An example would be the relevance of reiterative daylight analysis in the early design of a school building where, among other design criteria, desired daylight autonomy (DA) levels will help drive design decisions.

Building envelopes represent the primary node of confluence between design and performance – this is the transitionary zone for energy transfer between external and internal environments, and also a primary medium of external design expression. Beyond energy and aesthetics, the envelope design also contributes to the thermal, visual and acoustic comfort of a building's occupants. Given such a wide-ranging impact, it is imperative that a building envelope be designed with multi-variable goals.

Brigade WTC project
Chennai

Situated in Chennai in India, the 'World Trade Centre'-branded project for Brigade Group is a roughly 232,000m^2 mixed-use development on a 6.5-hectare site. This project includes residential, commercial, retail and hospitality components. Two commercial towers with a total area of roughly 150,000m^2 are key architectural features in the composition of the commercial development (see Figures 3, 4, 5 and 6). Given the size and prominence of these two towers, a great effort was put into designing their facades – not just for aesthetic results, but equally for the performance aspect of the towers.

Running a preliminary or 'box-model' energy analysis provided essential understanding of peak energy loads in the building. Box-models are simple enough and were run internally within the firm, but can also be required by the energy model consultant. They are literally boxed models with presumed building envelope, systems, occupancy and other internal loads. The model is then run in the project's climate zone to extract peak loads. Displaying this extracted data in a pie-chart format shows the breakdown of peak energy loads in percentage terms, and when compared side-by-side with a varying climate zone or program, this makes clear how loads shift. This type of data begins to form architecture that is now appropriately responsive to its context. A designer can take this information and prioritise design responses to address key energy loads on the building. As an example, a comparative energy load breakdown shown in Figure 7 shows an office building in three different cities: Abu Dhabi, Chennai and New York. Highlighted sections of the pie are loads attributable to the building envelope, which vary across climate zones even when the internal program is identical.

Furthermore, the breakdown of envelope loads reveals that in different climates these differ in terms of thermodynamics, and on whether solar gains are conductive or direct. Such information clarifies how a building envelope should be designed for Abu Dhabi versus Chennai, even though both are considered hot climates.

It is clear from the Chennai model that direct and indirect solar radiation is the primary load on building envelope. That is a clear clue to design a facade that prevents incident radiation on its fenestration, thus reducing energy transmittance through the glazing. A design approach to prevent solar radiation therefore takes precedence over insulation properties of the facade.

Figure 3:
Location map

Figure 4:
Site plan

Legend:
1 Civic stage
2 Landform seating
3 Reflecting pool
4 Urban bosque
5 Waterside lawn
6 Outdoor dining space
7 Arrival green corridor
8 Event lawn
9 Multi-sport court
10 Arrival pond
11 Jogging track
12 Urban dunes
13 Urban marsh
14 Buffer streetscape

Figure 5:
Elevation

112   Energy Modelling in Architecture

Figure 6:
Peak load breakdown

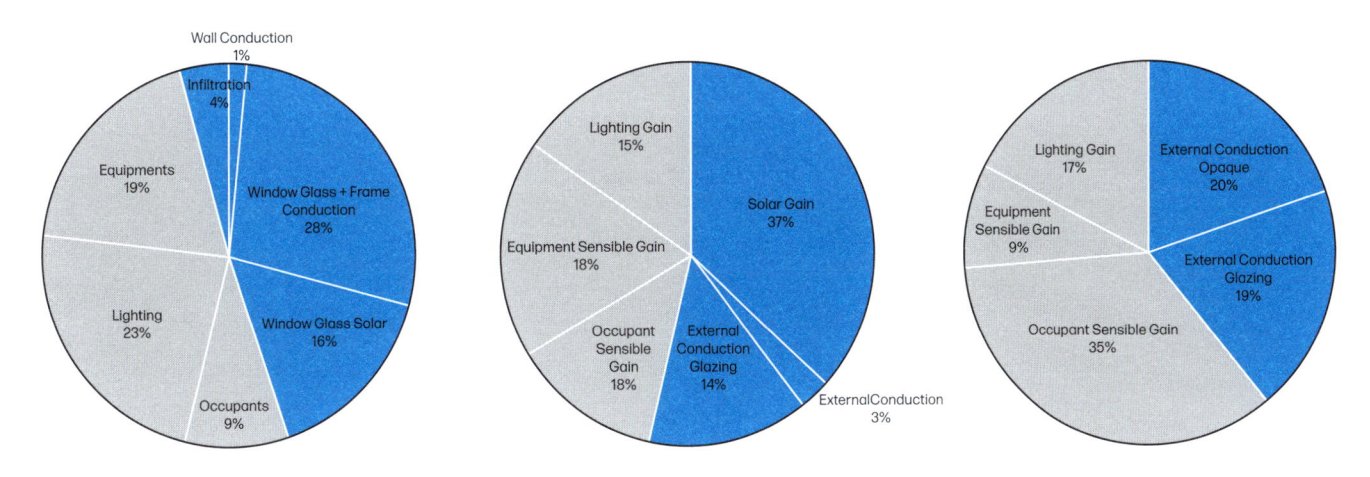

Figure 7:
Comparative energy load breakdown

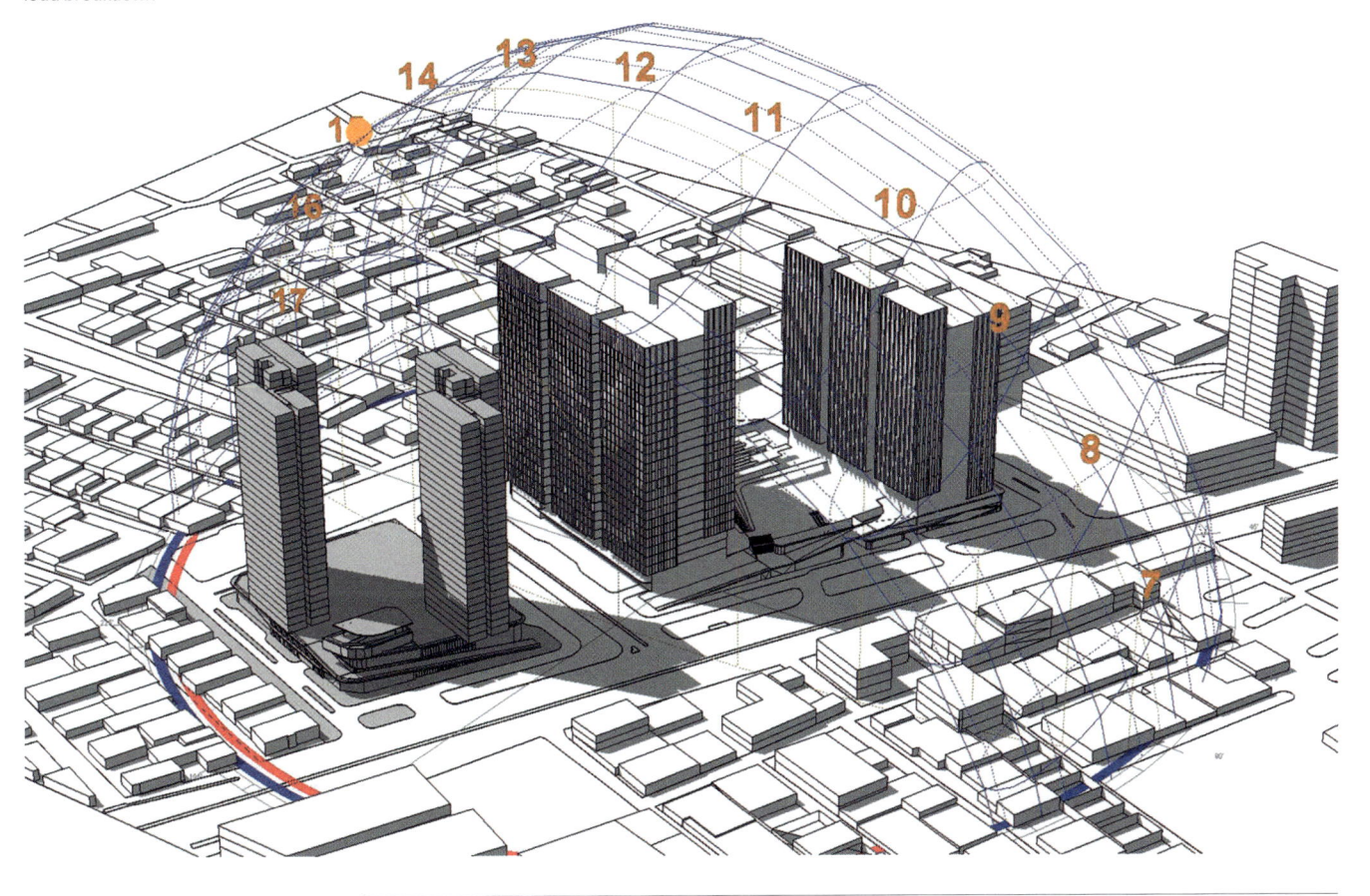

HOK - achieving contextual design through a measured process 113

Form sculpting

At the start of the design process, as the overall site plan was being designed, it was ensured that the commercial building footprint was maintained at a minimal width while oriented to maximise facade areas facing north and south. This would facilitate an easier design solution for preventing insolation transmittance through its skin, and better daylit spaces. Once the building massing and location were set and maximum zoning height achieved, the project team began work to design the articulation of the building massing and facade. With an understanding that radiation causes significant loads on this building, the goal was always to achieve a design expression that met the aesthetic goals while minimising exposure to direct and indirect solar radiation.

Of the two commercial towers, tower A-1, the taller of the two, is designed to be approximately 112m tall and is poised to be the tallest commercial tower in the city.

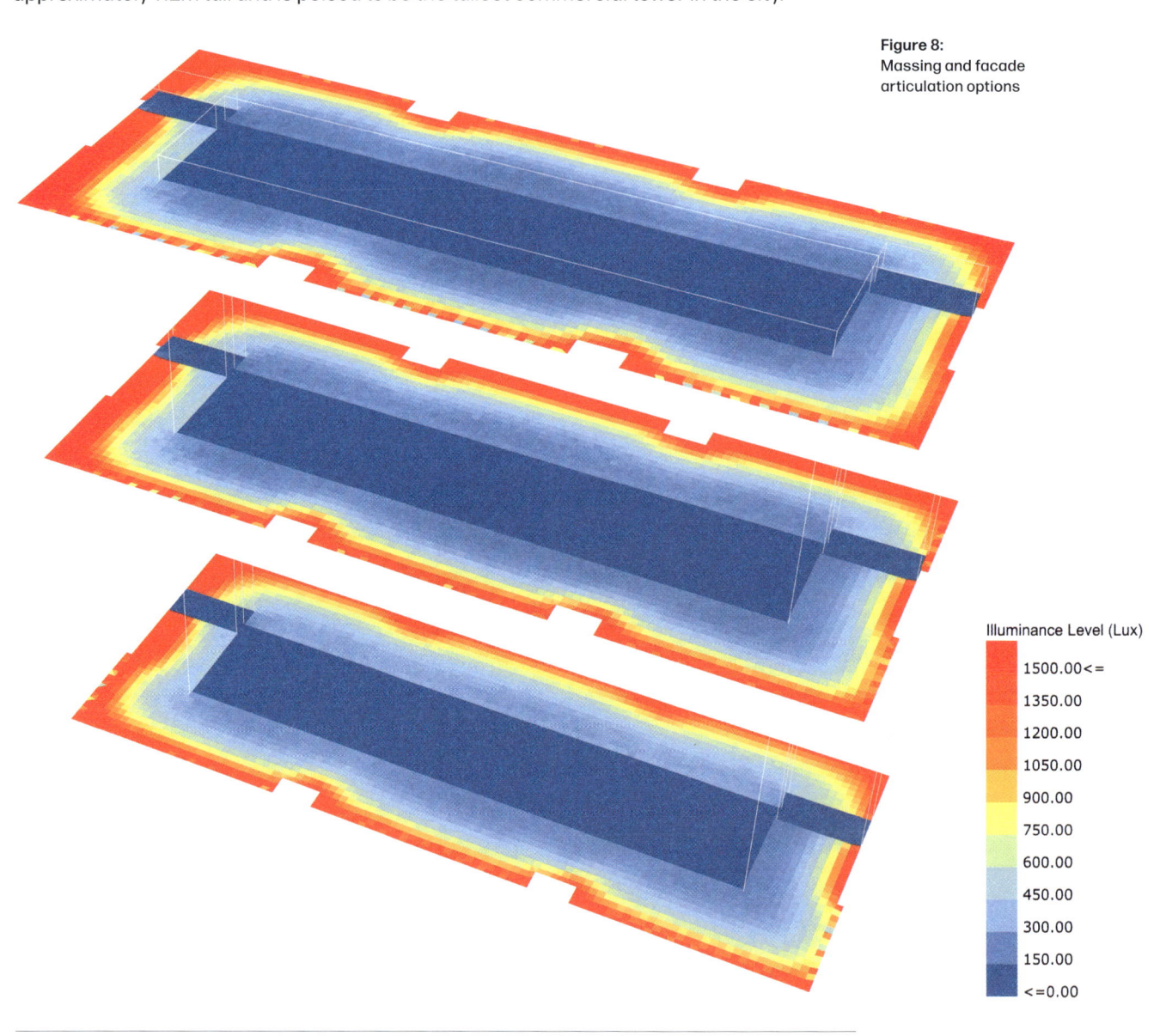

Figure 8:
Massing and facade articulation options

Illuminance Level (Lux)

1500.00<=
1350.00
1200.00
1050.00
900.00
750.00
600.00
450.00
300.00
150.00
<=0.00

114   Energy Modelling in Architecture

Accommodating a narrow floorplate to maximise floor area ratio (FAR) within a 112m-high tower means that the floorplates are fairly long, measuring roughly 85m. An extruded mass using such a floorplate results in a bar form more than a tower. To avoid this, the massing was broken up to read as a series of vertical towers. The design exercise to study the tower massing was a methodical process in which as a first step various configurations of massing divisions were studied. Varying massing strategies were then coupled with facade articulation strategies. As a third step in the design process, select massings with applied facade articulation were assessed for incident solar radiation. It should be noted that even while developing various options for massing and facade articulation, a play of depth was always kept intrinsic to create self-shading.

At the conclusion of this design exercise and in the process of evaluating the final schemes, an assessment of reduction in solar radiation was made in conjunction with the aesthetics of its massing and facade (see Figure 8). These studies and process diagrams were shared with the client during the concept design phase. Even though the client was keener on the aesthetic outcome of the massing studies, it was clear that they appreciated and took into account radiation analysis while evaluating the options.

Figure 9:
Facade design processes

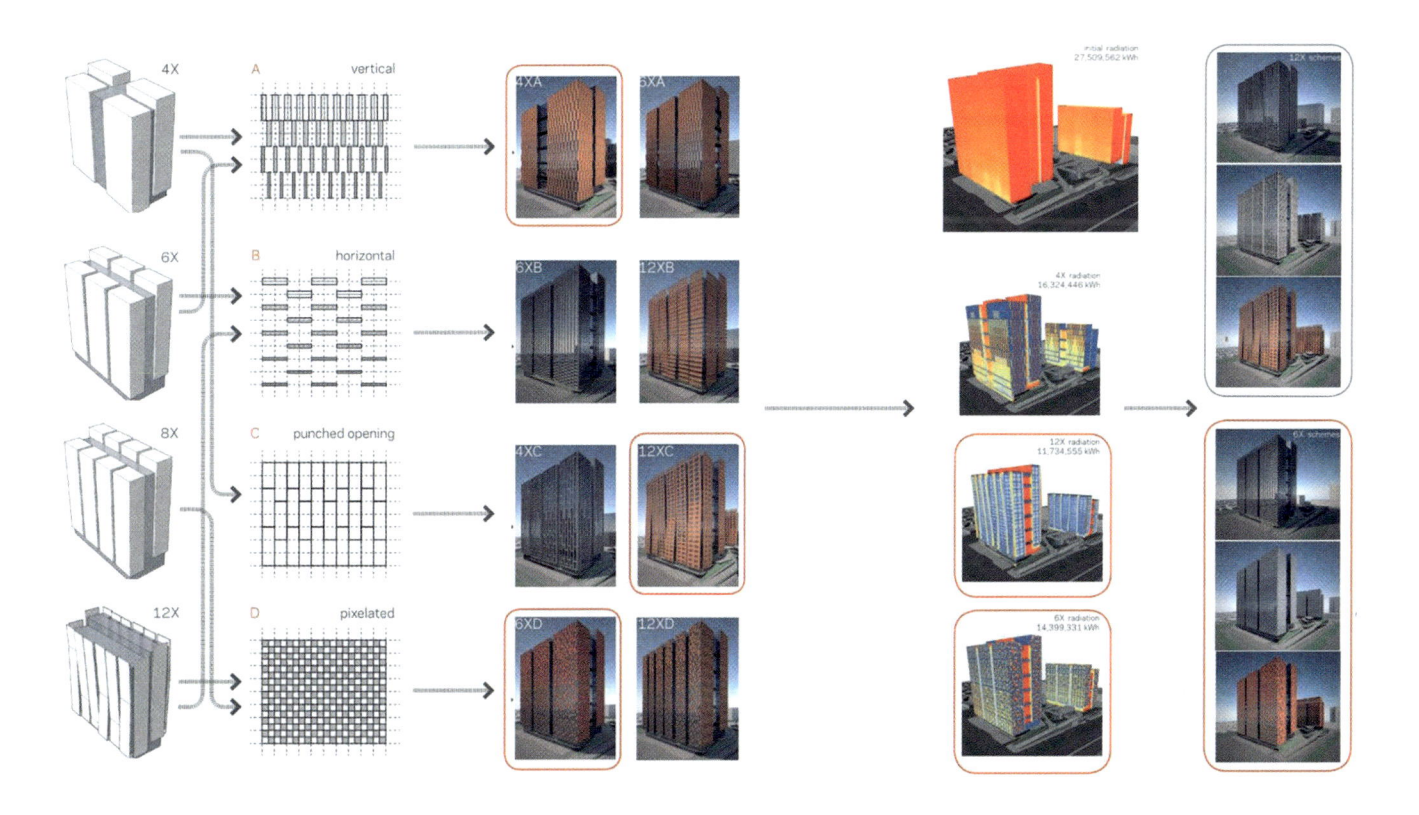

PANEL TYPE GENERATION

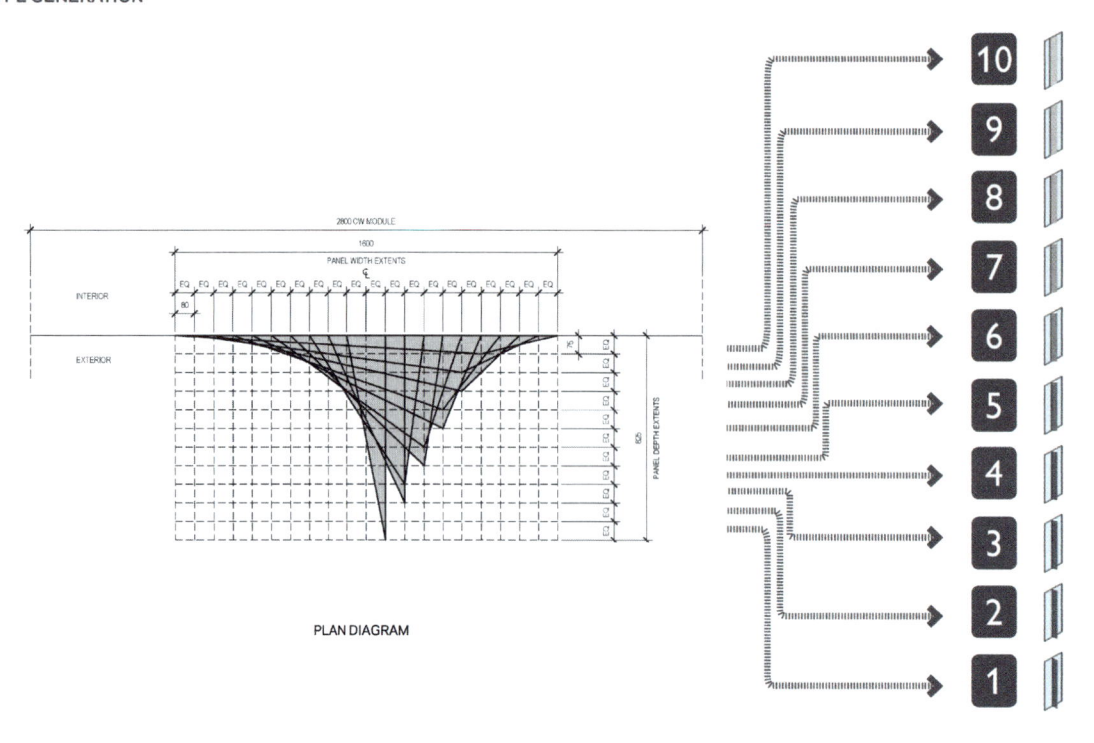

PLAN DIAGRAM

LEVEL 1 - INSTANTIATED PANEL TYPES

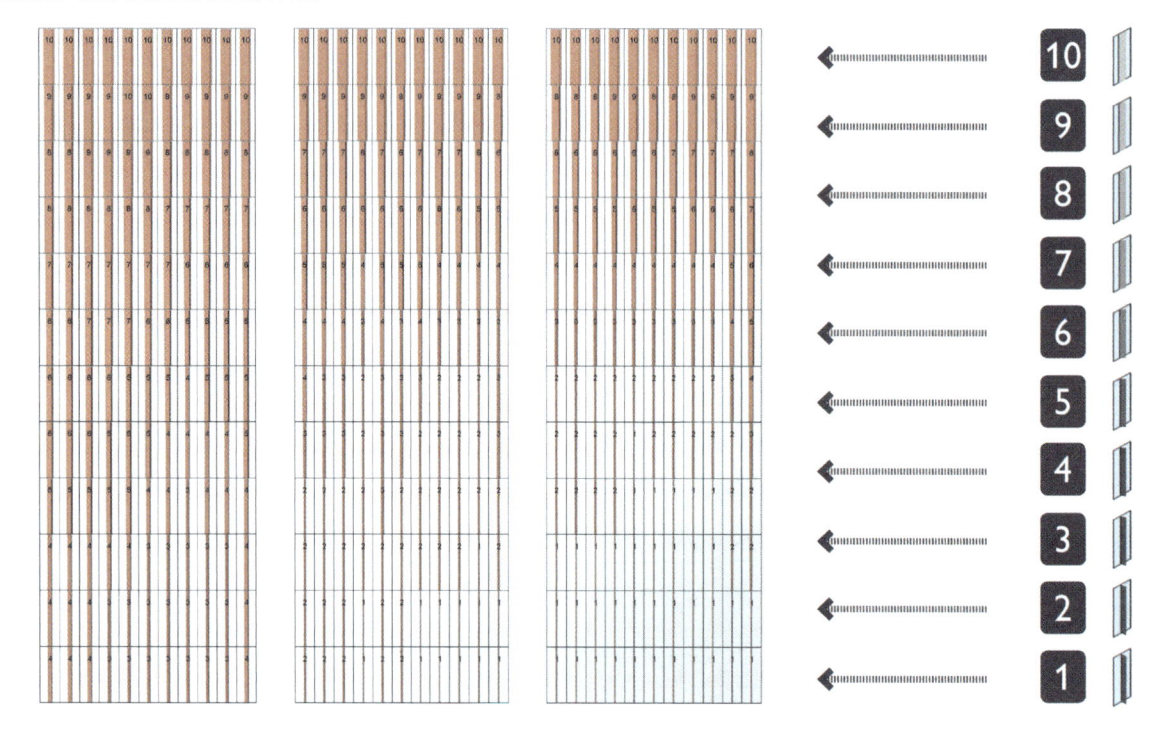

Figure 10:
Facade design processes

116 Energy Modelling in Architecture

LEVEL 1 - RADIATION ANALYSIS

Range	Level
660–700 kwh/m2	10
620–660 kwh/m2	9
580–620 kwh/m2	8
540–580 kwh/m2	7
500–540 kwh/m2	6
460–500 kwh/m2	5
420–460 kwh/m2	4
380–420 kwh/m2	3
340–380 kwh/m2	2
300–340 kwh/m2	1

FINAL DESIGN COMPARISON

Initial Radiation
13,878,286.85 kWh

Final Radiation
9,500,048.50 kWh

Reduction in annual incident radiation:

31.54%

HOK - achieving contextual design through a measured process

An environmental parametric skin

Once the tower massing was more or less finalised with the client, the team turned its focus to developing a design for its skin that carried the design idea of a vertical volume while ensuring that it responded to its external environment. The process employed to refine the envelope design was a combination of aesthetics, thermal analytics and geometry. The preliminary vertical fins were transformed into a triangular geometry that morphed from a wide and almost flattened panel form to a narrow-width yet deep fin profile in incremental steps. The variable width and depth created ten unique fins or panels that could now be applied to the facade.

At the same time, the facade was broken down into modules to enable construction of a unitised curtain-wall system. The same division pattern grid was also used to analyse incident solar radiation on the facade surface. A parametric definition created a rule by which a panel type was paired with the amount of incident solar radiation, as described in Figure 10. This parametric definition or script then generated a pattern on the entire facade based on solar radiation and geometry parameters defined by the design team. Such parameters can be altered to achieve a more desirable aesthetic outcome, or to further adjust the reduction of incident solar radiation.

A further analysis was run to assess solar transmittance into the building for all ten unique modules, and the results were intriguing. It was found that for every module, while the depth and width of the opaque fins changed, the amount of solar transmittance remained relatively constant for all facade orientations. Clearly, the strategy of deeper shading fins for larger glazed areas and shallower panels adjacent to smaller glazed areas seemed to balance out the solar transmittance through the fenestration. The result gave the design team more flexibility to achieve a more desirable design, knowing that the solar radiation metrics would remain almost constant. It also enabled the design to emphasise views via larger glazed areas on higher floors on the east side of the tower facing the sea, about 2km away. With all these design and performance metrics balanced and optimised, the design team were able to reduce incident radiation on the building facade by over 30%. Through whole-building energy analysis, HOK's designers realised savings of approximately 16%, attributable simply to building massing and facade design.

Conclusions

5

Recommendations and conclusion

The preceding chapters make it clear that the challenges facing larger practices of integrating energy modelling early in design are often quite different to those faced by smaller practices.

It seems that the challenges and tensions for large practices are largely focused on social and organisational leadership issues, whereas in small practices it tends to be about the relationship between the client and the designer. Medium-sized firms, on the other hand, are presented with challenges of evolving an experimental approach to developing shared workflows, and in-house ways of working that use other disciplinary skills and knowledge.

In all the discussions, there is a shared recognition of a collaborative multidisciplinary language that better weaves together the analytic and the creative, the unmeasurable and the measurable, the technical and the social, the architect and the engineer. For large firms, the continual maintenance of a knowledge/skill base within a practice where employees may not always be personally invested in environmental design/energy modelling best practice is highlighted. Active supervision and an atmosphere of sustainability are observed by FCBStudios, whereas HOK note the importance of sustainability leaders across offices and projects. AHMM discuss the important role in-house expertise can play in early optimisation of design approaches. Meanwhile, medium-sized firms emphasise the need for efficient flow of information between different consultants to enable an effective iterative design workflow. KieranTimberlake and Henning Larsen discuss the ways in-house expertise is developed through research and iterative approaches to refining workflows that take into consideration the social context and analytical boundaries of an energy model. Architype reflect on the value of PHPP and using energy modelling to define a set of design rules of thumb.

Rules of thumb and an individual approach to energy modelling built on detailed monitoring of built projects is also advocated by smaller firms bere:architects and Prewett Bizley. The importance of embodied energy and resource efficiency are discussed by Tonkin Liu in relation to their shell lace structure technology.

In all chapters, defining where particular responsibilities and liabilities lie is continually appraised and reappraised across projects, processes and practice. A common theme across all discussions is understanding the boundary of the energy model, particularly in terms of how far contextual issues may be considered. The contributions in this book offer a detailed account of how some of the early-adopter firms from across the world have approached this question, reflecting on ways in which the practices consider the integration of energy modelling in their projects.

The book presents hopeful insights into best practice that involves the production, evaluation and application of energy modelling in architecture. What we have found is that there are many knowledge practices engaged in this integration of energy modelling along the way that involve and depend on the social and organisational as well as the technological skills, knowledge and values within each firm.

References

1. Zero Carbon Hub, 'Closing the gap between designed and as built performance', Interim Progress Report, 2013.
2. S. Oliveira, E. Marco, B. Gething and S. Organ, 'Evolutionary, Not Revolutionary – Logics of Early Design Energy Modelling Adoption in UK Architecture Practice', *Architectural Engineering and Design Management*, vol. 13, no. 3, 2017, pp 168-184.
3. S. Oliveira, E. Marco, B. Gething and S. Organ, 'Evolutionary, Not Revolutionary – Logics of Early Design Energy Modelling Adoption in UK Architecture Practice', *Architectural Engineering and Design Management*, vol. 13, no. 3, 2017, pp 168-184.
4. S. Oliveira, E. Marco, B. Gething and S. Organ, 'Evolutionary, Not Revolutionary – Logics of Early Design Energy Modelling Adoption in UK Architecture Practice', *Architectural Engineering and Design Management*, vol. 13, no. 3, 2017, pp 168-184.
5. D.R. Cotton, W. Miller, J. Winter, I. Bailey and S. Sterling, 'Developing Students' Energy Literacy in Higher Education', *International Journal of Sustainability in Higher Education,* vol. 16, no. 4, 2015, pp 456–473.
6. S. Altomonte, P. Rutherford and R. Wilson, 'Mapping the Way Forward: Education for Sustainability in Architecture and Urban Design', *Corporate Social Responsibility and Environmental Management*, vol. 21, no. 3, 2014, pp 143–154.
7. S. Oliveira, E. Marco and B. Gething, 'Energy-Efficient Design and Sustainable Development', in W. Leal Filho (ed.), Encyclopedia of Sustainability in Higher Education, Springer, 2019.
8. Rocky Mountain Institute, 'Building Energy Modeling For Owners and Managers: A Guide to Specifying and Securing Services', https://rmi.org/wpcontent/uploads/2017/05/RMI_Document_Repository_Public-Reprts_2013-17_BUILDINGENERGYMODELINGFOROWNERSANDMANAGERS.pdf, 2013, (accessed 27th March 2020).
9. American Institute of Architects, 'An Architect's Guide to Integrating Energy Modeling in the Design Process', <http://content.aia.org/sites/default/files/2016-04/Energy-Modeling-Design-Process-Guide.pdf>, 2012, (accessed 27th March 2020).
10. E. Fasoulaki, 'Integrated Design: A Generative Multi-Performative Design Approach', Massachusetts Institute of Technology, 2008, p 7.
11. F.L. Flagler and J. Haymaker, 'A Comparison of Multidisciplinary Design, Analysis and Optimization Processes in the Building Construction and Aerospace Industries, Conference Paper', p 625.
12. F.L. Flagler and J. Haymaker, 'A Comparison of Multidisciplinary Design, Analysis and Optimization Processes in the Building Construction and Aerospace Industries', Conference Paper, p 626.
13. X. Li, et al., 'Building Energy Efficiency and Healthy Indoor Environment. Advances in Mechanical Engineering', 2014, p 190
14. J. Haymaker, et al., 'Design Space Construction: A Framework to Support Collaborative, Parametric Decision Making', *Electronic Journal of Information Technology in Construction*, 2018, p 158.
15. Welch, et al., *A BIM Workflow*, p 1.
16. W. Tian, et al., 'A Review of Uncertainty Analysis in Building Energy Assessment', *Renewable and Sustainable Energy Reviews*, 2018, p 286.
17. P. Clegg, *New Low-Cost Sources of Energy for the Home*, Storey Books, 1977.
18. https://www.oneplanetnetwork.org/initiative/one-planet-living-framework (accessed 20 January 2020).
19. https://www.theenvironmentalhandbook.com/ (accessed 20 January 2020).

Index

Page numbers in **bold** indicate figures.

A

active chilled beam strategy 58–59, **60**
Allford Hall Monaghan Morris (AHMM) 83–95, **84**, **86**
    compliance 94–95, **95**
    flexible office design 86–87, **86**, **87**, **88**
    solar gain modelling 89–92, **89**, **90**, **91**, **92**
    thermal mass modelling 92–94, **93**, **94**
American Institute of Architects (AIA) 108
architectural shading 40–41, **42**
    facade articulation strategies 110, **113**, 114–115, **114**, **115**, **116**, **117**, 118
Architype 65–79, **66**
    Bicester Eco Business Centre 73–79, **73**, **74**, **75**, **76**, **77**, **78**
    Ysgol Bro Hyddgen, Machynlleth 67–72, **67**, **68**, **69**, **70**, **71**, **72**
Arup 27, 29, **32**
Association for Environment Conscious Building (AECB) 17
Atelier Ten 42

B

bere:architects 21–25, **21**, **22**, **23**, **24**
Bexhill Pavilion 27, 28, **29**
Bicester Eco Business Centre 73–79, **73**, **74**, **75**, **76**, **77**, **78**
BIM software 45, 65, 68, 69, 72
box-models 110
Brigade WTC project, Chennai, India 110–118, **111**, **112**, **113**, **114**, **115**, **116**, **117**
British Council for Offices (BCO) 89
Broadcasting Place, Leeds Metropolitan University 101, **102**
Building Regulations
    Denmark 51, **51**, 53, **54**
    UK 21, 83, 94
BuroHappold 39, 56, 62

C

Carl H. Lindner College of Business, University of Cincinnati 56–60, **57**, **58**, **60**, **61**, 62
CIBSE TM52 95, **95**
computational fluid dynamics (CFD) modelling 56, 59, 60
cooling
    active chilled beam strategy 58–59, 60
    thermal mass modelling 92–94, **93**, **94**
    vegetative modelling 38–40, **39**, **40**
    see also shading

D

Danish energy performance framework 51, **51**, 53, **54**
data science methods 46
daylight modelling 56, 60, **61**, 99–101, **100**, **102**, 109
    see also solar gain
dynamic thermal modelling 56, 59, 60, **60**

E

electrical submetering 21–25, **21**, **22**, **24**
embodied energy, shell lace structures 31
Energy for the Home 97, **97**
energy reduction strategies 52–53, **53**, **54**
external shading devices 19, 75–76, **75**, **76**, 91–92, **91**, **92**

F

facade articulation strategies 110, **113**, 114–115, **114**, **115**, **116**, **117**, 118
feedback loops *see* KieranTimberlake
Feilden Clegg Bradley (FCB) Studios 7, 97–104, **98**, **100**, **102**, **103**
form factor comparison 14, **14**

G

glazing performance 41, **42**, 58, 59, **60**

H

Henning Larsen 7, 51–62
    Carl H. Lindner College of Business, University of Cincinnati 57–60, **57**, **58**, **60**, **61**, 62
    integrated energy design 52–56, **52**, **53**, **54**
    Kolding Campus, University of Southern Denmark 53, **54**
    praxis and workflow 55–56, **55**
higher education, interdisciplinary learning 9, **9**
Hive library, University of Worcester 101, **103**
HOK 102–118
    Brigade WTC project, Chennai, India 110–118, **111**, **112**, **113**, **114**, **115**, **116**, **117**
    guidelines and processes 108, **109**
    sustainable design leaders 107, **107**
hygrothermal performance 65

I

integrated energy design 52–56, **52**, **53**, **54**
internal wall insulation 15, **16**

K

Kahn, Louis 107
KieranTimberlake 9, 37–48, **43**, **46**
    architectural shading 40–41, **42**
    technical and social challenges 44–48, **48**
    vegetative modelling 38–40, **39**, **40**
Kolding Campus, University of Southern Denmark 53, **54**

L

landscape performance studies 38–40, **39**, **40**
Lark Rise, Buckinghamshire **21**, **22**, 23–25, **23**, **24**
learning approaches 7–9, **8**, **9**
lighting power density (LPD) analysis 108

M

machine learning 46
Manchester Tower of Light 31, **32**, **33**
massing and facade articulation strategies 110, **113**, 114–115, **114**, **115**, **116**, **117**, 118
multidisciplinary approach *see* Henning Larsen; KieranTimberlake

N

nature-informed design 27–32, **27**, **28**, **29**, **30**, **32**, **33**

O

One Planet Living framework 98–99, **98**, 104
overheating risks 16, **17**, 19, **57**, 75, 100–101, **102**
    see also solar gain

Index   127

P

passive energy reduction strategies 52-53, **53**, **54**
Passive House 65
    post-occupancy evaluation (POE) 19, 21-25, **21**, **22**, **24**
    retrofit projects 13-19, **14**, **15**, **16**, **17**, **18**, **19**
Passive House Planning Package (PHPP) 13-19, **14**, **15**, **16**, **17**, **18**, **19**, 24-25, **24**
    see also Architype
Philadelphia Navy Yard 39, **39**, **40**
physics-based models 44-45
post-occupancy evaluation (POE) 19, 21-25, **21**, **22**, **24**, 48, 85
Prewett Bizley 7, 13-19, **14**, **15**, **16**, **17**, **18**, **19**

R

resource-efficiency first approach 27-32, **27**, **28**, **29**, **30**, **32**, **33**
retrofit projects, Passive House 13-19, **14**, **15**, **16**, **17**, **18**, **19**
rules of thumb 102-103

S

Sefaira software 7, **8**, 100, **100**
sensor data 48
shading
    architectural shading 40-41, **42**
    external shading devices 19, 75-76, **75**, **76**, 91-92, **91**, **92**
    facade articulation strategies 110, **113**, 114-115, **114**, **115**, **116**, **117**, 118
    vegetative modelling 38-40, **39**, **40**
shared workflows *see* Henning Larsen; KieranTimberlake
shell lace structures 27-32, **27**, **28**, **29**, **30**, **32**, **33**
SketchUp software 99-100
smart controls 25
social challenges 44-48, **46**
software 45, 48, 55, **55**, 65, 85, 101, 103
    BIM 45, 65, 68, 69, 72
    Sefaira 7, **8**, 100, **100**
    SketchUp 99-100
    see also Passive House Planning Package (PHPP)
solar gain 100-101
    architectural shading 40-41, **42**
    external shading devices 19, 75-76, **75**, **76**, 91-92, **91**, **92**
    massing and facade articulation strategies 110, **113**, 114-115, **114**, **115**, **116**, **117**, 118
    office design projects 89-92, **89**, **90**, **91**, **92**
    university building projects **57**, **58**, 59, **60**, **61**, 101, **102**
    vegetative modelling 38-40, **39**, **40**
Standard Assessment Procedure (SAP) 13
sustainability leadership 107, **107**
sustainability reviews 98-99, 104
system boundaries 38-44, **39**, **40**, **42**

T

Technology Strategy Board (TSB) 21
thermal bridging 15, **16**, 17, 19, 65, **72**, 77, **77**
thermal mass modelling 92-94, **93**, **94**
    see also massing and facade articulation strategies
Tonkin Liu 27-32, **27**, **28**, **29**, **30**, **32**, **33**

V

vegetative modelling 38-40, **39**, **40**

W

window-to-wall ratios 57, **58**, 59, **61**

Y

Ysgol Bro Hyddgen, Machynlleth 67-72, **67**, **68**, **69**, **70**, **71**, **72**

Image credits

Introduction	Sonja Oliveira
Chapter 1	All images Sonja Oliveria
Chapter 2	All images Prewett Bizley
Chapter 3	All images Justin Bere
Chapter 4	Figures 1 and 5: Tonkin Liu / Arup, all other images Tonkin Liu
Chapter 5	All images KieranTimberlake
Chapter 6	All images Henning Larsen
Chapter 7	All images Architype
Chapter 8	All images AHMM
Chapter 9	Figure 2: Bioregional, all other images Feilden Clegg Bradley Studios
Chapter 10	All images HOK